Green Etching Techniques for MEMS Applications

Green Etching Techniques for MEMS Applications: Sustainable, Fluorine-Free Etching Methods for Micro-Electro-Mechanical Systems delivers an essential and comprehensive exploration of sustainable, fluorine-free etching methodologies for microelectromechanical systems (MEMS). With growing environmental concerns around traditional MEMS fabrication, this book addresses a critical issue by detailing advanced, eco-friendly alternatives that mitigate environmental impacts without compromising technical performance. Covering a spectrum of innovative etching technologies, from dry and wet chemistries to electrochemical and AI-enhanced hybrid methods, the book offers practical guidance, robust theoretical foundations, and insights from real-world industrial case studies. It is a crucial resource for professionals, researchers, and students dedicated to advancing the sustainability of MEMS fabrication.

Features:

- Thorough analysis of fluorine-free, environmentally friendly MEMS etching alternatives
- Coverage of emerging technologies, including supercritical CO_2, ionic liquids, and gas-phase selective etching
- Exploration of AI-driven process optimization for sustainable and efficient MEMS manufacturing
- Detailed industrial case studies highlighting successful implementation and scalability
- Clear discussions on regulatory drivers, market trends, and future roadmaps for sustainable microfabrication

Green Etching Techniques for MEMS Applications

Sustainable, Fluorine-Free Etching Methods for Micro-Electro-Mechanical Systems

Kaiying Wang

CRC Press
Taylor & Francis Group
Boca Raton London New York

CRC Press is an imprint of the
Taylor & Francis Group, an **Informa** business

Designed cover image: Shutterstock

First edition published 2026
by CRC Press
2385 NW Executive Center Drive, Suite 320, Boca Raton, FL 33431

and by CRC Press
4 Park Square, Milton Park, Abingdon, Oxon, OX14 4RN

CRC Press is an imprint of Taylor & Francis Group, LLC

© 2026 Kaiying Wang

ISBN: 978-1-041-14496-0 (hbk)
ISBN: 978-1-041-14509-7 (pbk)
ISBN: 978-1-003-67479-5 (ebk)

DOI: 10.1201/9781003674795

Typeset in Times
by KnowledgeWorks Global Ltd.

Contents

Preface

The accelerating demand for microelectromechanical systems (MEMS) across diverse industries—from healthcare and telecommunications to consumer electronics and automotive applications—has spurred continuous innovation in their design and fabrication. Yet, as we push the boundaries of miniaturization and performance, a critical challenge remains largely unresolved: the environmental and health impact of traditional MEMS etching techniques. Conventional plasma and wet etching processes rely heavily on hazardous chemicals and fluorinated gases, which pose significant risks to human safety, energy consumption, and global sustainability.

Green Etching Techniques for MEMS Applications was conceived to address this gap. This book aims to provide a comprehensive, structured, and accessible guide to the evolving field of sustainable MEMS etching. It introduces readers to a wide array of fluorine-free, energy-efficient, and environmentally friendly etching alternatives, encompassing dry and wet etching, electrochemical strategies, and AI-enhanced process control. From foundational theories to advanced industrial case studies, this volume charts a path forward for researchers, engineers, and policy makers who are committed to reducing the ecological footprint of microfabrication.

The chapters in this book are designed to serve both as a practical reference and a conceptual roadmap. Readers will find rigorous discussions of the mechanisms underlying various etching techniques, detailed comparisons of fluorine-free alternatives, and insights into how AI and robotics can transform etching into a smart, sustainable process. Special attention is given to the challenges of scalability, material compatibility, and industrial adoption—topics that are critical for translating laboratory innovations into real-world impact.

Whether you are an academic researcher exploring next-generation fabrication methods, an engineer working to implement greener solutions in semiconductor manufacturing, or a student learning about MEMS technology for the first time, I hope this book serves as a valuable companion in your journey toward a more sustainable future for microfabrication.

Kaiying Wang

University of South-Eastern Norway,
Horten, Norway, July 10, 2025

Acknowledgments

We would like to express our deepest gratitude to the twenty-six EU Erasmus and USN master's degree students who took the course *SSI Design Lab II – Manufacturing Processes for Smart Systems (JMN455)* and *Microfabrication Technology (MFA4000)* at the University of South-Eastern Norway. Their hard work, dedication, and insightful contributions have been instrumental in the successful completion of this book.

The contributions of the students to the chapters are as follows:

- Chapter 1: Adnan, Md. Sahib Alam Provat, Md. Shadman Shakif Bhuiyan, Vaibhav Ashok Kumar
- Chapter 2: Gabriel Andreas Xavier Facusse Zepeda, Asma Mohiuddin, Hesham Mohamed Ellabban, Laura Daniela Gallego García
- Chapter 3: Fjolla Ahmetaj, Pragya Mallick, Hoàng Vũ Nguyễn
- Chapter 4: Ikenna Idu Henry, Ismael Musa, Maira Ehsan Mughal, Natnael Masresha Zerihun
- Chapter 5: Sanket Sandeep Deshmane, Ali Ghasemibousjin, Džeina Dombrava, Pegah Haghshomar
- Chapter 6: Nico Wichmann, Rodrigo Zamagno Medeiros, Lina Khaled Alkerdi, Muhammad Ukasha
- Chapter 7: Amir Abdelkareem Osman Tahir, Ahmed Abdullah Nasir, Mohammad Toufan, Ziwei Liu

Their enthusiasm and commitment to this project have been truly commendable. We are grateful for their valuable insights and volunteer contributions, which have significantly enriched the quality and depth of this book.

I would also like to extend my sincere appreciation to all the highly intelligent students from around the world who joined the programme and studied at the University of South-Eastern Norway as part of their mandatory course. Your hard work, innovative thinking, and global perspectives have brought immense value to the learning environment and have played a significant role in the success of this book project.

As the main professor responsible for this course, I would like to give my highest appraisal to these students for their hard work and their very positive attitude throughout the course. Their professionalism, dedication, and willingness to engage deeply with the subject matter have made this project a remarkable success.

Special thanks also go to the academic and technical staff at the University of South-Eastern Norway for their guidance and support throughout the development of this book.

This book is a testament to the spirit of collaboration and shared knowledge. Thank you to everyone who contributed to making this work a success.

Kaiying Wang

University of South-Eastern Norway

Author

Kaiying Wang received his PhD in condensed matter physics from the Institute of Physics, Chinese Academy of Sciences. He joined the University of South-Eastern Norway (USN) in 2007 as an associate professor and was promoted to professor in 2010. His research interests focus on micro-fabrication and nanotechnology, electrochemistry, photochemistry, and nanodevices for environment and energy applications. For teaching, he has taught Microfabrication (MFA4000, master) and Nanotechnology (TSE3120, bachelor) at the University of South-Eastern Norway since 2010. Meanwhile, he has participated in several national and European projects related to micro/nano fabrication technology. The related education programmes at USN include: (1) Sensor Systems and Innovation, (2) Smart Systems Integrated Solutions, and (3) Micro- and Nano Systems Technology.

1 Introduction to Green Etching in MEMS Manufacturing

1.1 ENVIRONMENTAL IMPACT OF TRADITIONAL MEMS ETCHING

Microelectromechanical systems (MEMS) fabrication relies on a sequence of deposition, lithography, and etching steps to carve microscopic structures. While MEMS devices are physically small, their manufacturing can carry a surprisingly large environmental footprint. The use of harsh chemicals, plasma gases, and energy-intensive equipment in traditional etching processes leads to several ecological concerns [1, 2].

This section discusses the key environmental impacts associated with conventional MEMS etching, structured into three focus areas: (1) the emission of fluorinated greenhouse gases from plasma etching, (2) the generation of hazardous chemical wastes in wet and dry etching, and (3) the high energy consumption of plasma-based etching systems like deep reactive ion etching (DRIE). Understanding these issues is a first step toward developing "green" etching techniques that mitigate harm to the environment.

Plasma etching in MEMS often employs fluorinated gases such as sulfur hexafluoride (SF_6), nitrogen trifluoride (NF_3), tetrafluoromethane (CF_4), and related perfluorocarbons. These gases are valued for producing reactive fluorine species that etch materials like silicon and silicon dioxide with high anisotropy and selectivity. For example, the Bosch DRIE process for silicon commonly alternates SF_6 (as the silicon etchant) with C_4F_8 (for sidewall passivation) [2]. CF_4 and other fluorocarbons are widely used in reactive ion etching (RIE) of oxides and nitrides, and NF_3 is often used for in-situ chamber cleaning processes. However, a serious drawback is that many of these compounds are potent greenhouse gases with extremely long atmospheric lifetimes. Any fraction that escapes the fab's abatement systems can accumulate in the atmosphere and contribute significantly to global warming [3].

The global warming potential (GWP) of fully fluorinated etch gases is enormous. SF_6, for instance, has a 100-year GWP on the order of 22,800–23,900 (meaning 1 kg of SF_6 has the warming effect of over 22,000 kg of CO_2) and an atmospheric lifetime of around 3200 years. Similarly, CF_4 persists for roughly 50,000 years in the atmosphere with a GWP in the range of 6500–7000 [3]. NF_3, a cleaning gas introduced as a "better" alternative to some perfluorocarbons, has a GWP on the order of 16,000 and lasts several centuries in the atmosphere [4]. These values far exceed those of CO_2 and even most other industrial gases. Although the semiconductor industry's absolute emissions of such gases are small compared to CO_2, their outsized radiative impact makes them significant climate risks [3]. Prather and Hsu famously dubbed NF_3 "the greenhouse gas missing from Kyoto," highlighting that it was unregulated for years despite its high GWP [4, 5].

DOI: 10.1201/9781003674795-1

The use of SF_6 and related gases in MEMS manufacturing has grown alongside the expansion of the MEMS market, raising concerns about increasing emissions. In deep silicon etching, large volumes of SF_6 and C_4F_8 are expended to achieve high aspect ratio structures; without proper abatement, these can be released to the atmosphere [6]. One industry study noted that the greenhouse gas output from SF_6/C_4F_8-based silicon etching was rising measurably with MEMS production and called for process changes to curb emissions. The semiconductor sector has responded over the past two decades by implementing emission reduction strategies, including plasma scrubbers and alternative chemistries [7]. For example, remote plasma cleaners using NF_3 can decompose more completely, and new etch gases (like unsaturated fluorocarbons) are being explored for lower GWP profiles [6]. Nonetheless, even with abatement technology, a portion of these fluorinated gases often escape. It is estimated that on average 10% or more of the fluorocarbon input can remain unreacted and be emitted as process by-products like CF_4 during etching and cleaning [3]. The result is that plasma etching and cleaning processes are non-negligible contributors to the carbon footprint of chip fabrication. In a life-cycle assessment of a MEMS device, fluorinated etch gases were found to be major contributors to the overall carbon emissions (measured in CO_2-equivalent), outweighing many other process inputs due to their high GWPs [2]. Reducing or replacing these gases is therefore a priority in "green" MEMS processing. Until such solutions mature, every deep etch using SF_6 or clean cycle using NF_3 links even the tiniest MEMS device to global climate change in a disproportionate way.

Beyond gaseous emissions, traditional etching processes generate significant chemical waste. Wet etching is widely used in MEMS for processes such as silicon substrate thinning (e.g., Potassium Hydroxide (KOH) etching) or releasing sacrificial layers (e.g., hydrofluoric acid (HF) etching of SiO_2). These wet etchants are often strong acids and bases that become hazardous waste after use. A modern semiconductor fab uses over a hundred different high-purity chemicals, and many of them are used in etching and subsequent cleaning steps. The spent solutions contain not only the original reagents but also dissolved metals, dopants, and reaction by-products which can be toxic. For instance, wet etching of silicon with KOH produces silicate-containing waste, and HF etching of oxide generates fluorosilicates. If inadequately treated, such effluent can contaminate water and soil, posing risks to ecosystems and human health [7].

HF is especially notable as an etchant in MEMS and semiconductor processing. HF is used to etch silicon dioxide and to clean wafer surfaces, and it is notoriously dangerous—capable of causing severe chemical burns and bone damage in humans. From an environmental standpoint, HF is one of the largest contributors to hazardous waste in chip fabrication. According to data from the Semiconductor Industry Association (SIA), waste solutions of HF account for over 40% of all hazardous waste generated by the semiconductor industry [7]. This startling figure reflects both the ubiquity of HF usage and the strict handling required for HF waste. Such waste acid cannot be simply flushed away; it must undergo neutralization and removal of fluoride ions (for example, by precipitation with calcium to form insoluble CaF_2) before disposal. Many fabs employ elaborate wastewater treatment facilities to manage HF and other acidic wastes, using methods like chemical coagulation, ion exchange, and reverse osmosis to recover or neutralize harmful chemicals [7]. Even so, accidents and leaks can pose environmental hazards, and the volume of waste acids remains a challenge as manufacturing scales up.

Dry etching processes, while avoiding liquid chemicals, produce their own types of waste. In plasma etching, the effluent gas stream can contain unreacted process gases (like SF_6, CF_4, Cl_2, etc.) as well as reaction by-products (e.g., SiF_4, C_xF_y polymers, and particulate matter). These gases cannot be released untreated; most fabs employ abatement systems such as burn boxes or scrubbers to decompose or capture toxic and greenhouse gases after the etch chamber. For example, perfluorocarbon gases may be combusted at high temperatures to break them into HF and CO_2, which are then neutralized and filtered [6, 7]. Abatement efficiency is high but not absolute—a fraction of perfluorocarbons (PFC)s and other gases still escape into the air. Moreover, the scrubbing process can generate secondary wastes: incineration of fluorinated gases produces HF vapor and fluoride salts that must be dealt with, and dry scrubbers eventually produce solid waste that needs disposal. Another waste issue in dry etching is the deposition of polymer residues. In processes like the Bosch DRIE, a fluorocarbon polymer (from C_4F_8) coats surfaces to protect sidewalls. This polymer must later be stripped off the wafer (usually by plasma O_2 ashing or solvents), and it also builds up inside chambers and vacuum lines. Periodic chamber cleans (often using gases like O_2, NF_3, or C_2F_6 plasmas) are required to remove these deposits, generating additional gaseous and particulate waste [3]. The removed polymer particles and spent chamber-clean effluent become part of the waste stream as well. All of these factors make clear that etching is a messy business: if not carefully managed, both wet and dry etch steps can output a variety of hazardous substances—from acidic liquids to toxic fumes—into the environment.

Handling and disposing of etching waste is therefore a critical environmental management issue for MEMS manufacturers. Modern facilities invest heavily in waste treatment infrastructure to comply with environmental regulations. For liquid wastes, this includes segregating streams (acids, bases, solvents, heavy-metal-bearing solutions) and treating each with appropriate chemical processes to neutralize toxins [7]. Solid hazardous wastes, like sludge from wastewater treatment or spent absorbents, must be disposed of as chemical waste at certified facilities. Despite these efforts, the sheer scale of production means significant waste generation. A recent analysis of the semiconductor foundry sector showed that hazardous waste output has been increasing in absolute terms, even as production processes become more advanced. Over the past decade, total hazardous waste from chip fabrication rose by over 200% in volume, reflecting the intense material usage of modern processes [8]. This trend underscores that without changes in process chemistry, scaling up fabrication will proportionally scale up waste. In summary, traditional etching techniques—both wet and dry—pose serious disposal challenges. With the expansion of the semiconductor industry, determining the huge associated water and energy consumption and accomplishing sustainable development can be key issues for this industry [9]. They require careful end-of-pipe management to prevent environmental contamination, and they motivate the search for cleaner alternatives that generate less harmful waste by design.

Etching processes, particularly plasma-based dry etching, are energy-intensive. They typically occur in vacuum chambers that use radio-frequency (RF) power to generate plasmas and require supporting systems like vacuum pumps, RF generators, heaters, chillers, and abatement units—all of which consume substantial electrical energy. DRIE, a cornerstone of MEMS fabrication for high-aspect-ratio features, exemplifies a high-energy process. DRIE systems employ powerful inductively coupled plasma sources (often running at kilowatts of RF power) and may run for extended durations (multiple hours of etch time for thick substrates) to achieve deep etch depths. Maintaining the plasma and the required low-pressure (on the order of milliTorr) over long cycles translates to significant electricity usage per wafer [2]. In essence, to etch deep trenches or release structures in MEMS, one must spend a lot of energy in the form of plasma power and vacuum operation.

Life-cycle analyses have highlighted the impact of this energy consumption. A recent bottom-up life cycle assessment (LCA) of a MEMS pressure sensor process found that the cumulative energy demand for fabricating a 1 cm² device was on the order of 0.7 kWh per cm² of wafer processed [2]. Figure 1.1 shows the energy consumption and water withdrawal of the semiconductor corporations in 2021. This includes the energy for all processing steps (deposition, lithography, etching, etc.) and the overhead of running a cleanroom. Notably, the single largest portion of fab energy use is often the operation of the cleanroom and equipment rather than chemical reactions themselves. Keeping air ultra-clean and dry, running pumps and thermal controls, and lighting/HVAC for the cleanroom contribute a constant background load. Within that, plasma

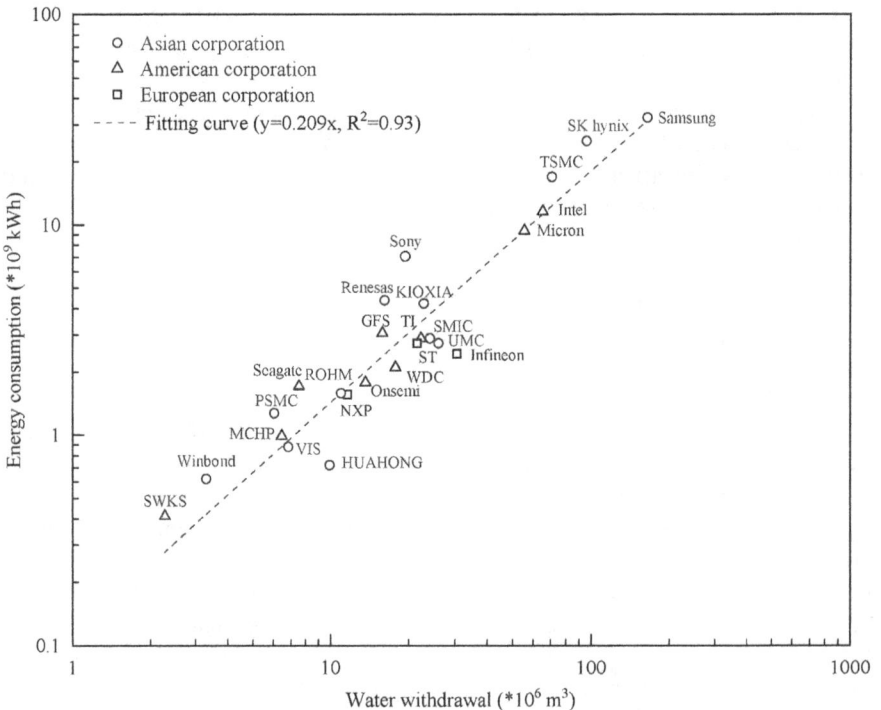

FIGURE 1.1 The energy consumption and water withdrawal of the semiconductor corporations in 2021 [9].

etching tools are among the more power-hungry pieces of equipment. For instance, an inductively coupled plasma (ICP) etcher for DRIE may require an RF generator of >1 kW, and turbopumps drawing additional hundreds of watts, all running continuously during the etch. If a process involves multiple long DRIE steps, the energy usage adds up correspondingly. Delhaye et al. compared two MEMS process flows and found that the one relying on lengthy DRIE steps had a noticeably higher carbon footprint attributed to energy and gas usage, compared to an alternative flow that used wet etching for certain steps [2]. In other words, an etch process that takes twice as long can roughly double the energy consumed by that tool, and if the electricity is generated from fossil fuels, the indirect CO_2 emissions will increase in proportion.

Another way to appreciate the energy intensity is to consider the embodied energy of the materials and infrastructure. Producing the ultra-pure silicon wafers used for MEMS already requires enormous energy input (on the order of 2000 kWh per kilogram of silicon). All subsequent processing adds to that initial "energy investment." During plasma etching, much of the energy is ultimately dissipated as heat (plasma heating of chamber walls, etc.), which must be removed by cooling systems, often leading to additional energy expenditure for refrigeration. Thus, plasma etching not only consumes energy directly but also loads the fab's cooling systems. The overall fab energy demand is significant: a single semiconductor manufacturing facility can consume tens of millions of kWh of electricity annually to support its processes. Studies have shown that the fabrication phase dominates the life-cycle energy and carbon impacts of electronic devices, far more than the use phase for small devices. Within that fabrication phase, processes like plasma etching are key contributors to the high electricity usage [10].

In the context of deep silicon etching for MEMS, the high energy cost is a trade-off accepted for technical capability. DRIE allows incredible precision and aspect ratios in silicon micromachining, but its "price" is paid in greenhouse gas emissions and energy bills unless mitigated. This has motivated research into more energy-efficient plasma sources and process optimization to shorten etch times. Some newer etch tools aim to lower RF power requirements through better plasma confinement or use alternative methods (e.g., cyclic etching that optimizes chemical efficiency) to remove material with less energy per unit depth [11]. Nonetheless, as of today, a typical MEMS manufacturer must reckon with the fact that every deep plasma etch step carries an environmental cost in kilowatt-hours. Reducing energy consumption in etching is part of the broader push for sustainable microfabrication, which also encompasses using renewable energy sources for fabs and designing processes that minimize time under power.

1.2 INDUSTRY SHIFT TOWARD SUSTAINABLE MEMS MANUFACTURING

The significant environmental footprint of traditional MEMS etching—characterized by greenhouse gas emissions, hazardous waste, and high energy consumption— has raised growing concerns among regulators, industry stakeholders, and the public. A recent study by Ruberti [12] highlights the substantial environmental burden of semiconductor fabrication, which shares many core processes with MEMS manufacturing. The average greenhouse gas emission intensity was found to exceed 10,000 kg CO_2 per square meter of wafer, largely due to energy-intensive etching,

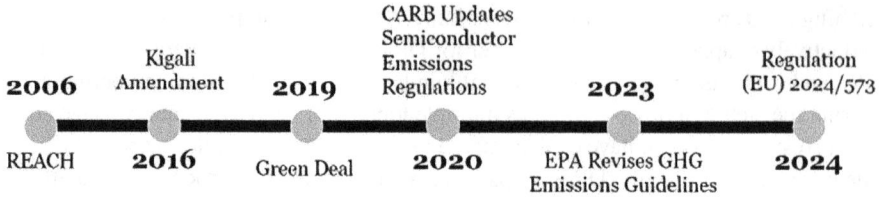

FIGURE 1.2 Timeline of major government and policy milestones.

cleaning, and deposition steps. In response, the MEMS industry is shifting toward greener practices, driven by evolving regulations and international climate policies targeting high-GWP gases and promoting sustainable manufacturing.

The shift toward sustainable MEMS manufacturing has been significantly influenced by evolving environmental regulations. Over the last decade, both international agreements and region-specific policies have pushed semiconductor and MEMS industries to reduce their reliance on high-GWP gases commonly used in etching processes [13]. These measures aim not only to reduce emissions but also to encourage cleaner alternatives across the microfabrication ecosystem. This timeline (Figure 1.2) highlights key regulatory milestones that have shaped this transition.

Regulatory pressure on MEMS manufacturing began with the EU's REACH regulation, restricting hazardous chemicals since 2006. The Kigali Amendment (2016) called for a global phasedown of HFCs, followed by the European Green Deal (2019), which set the path to net-zero emissions. Regionally, CARB (2020) and the EPA (2023) introduced stricter fluorinated gas controls in semiconductor fabs. Most recently, Regulation (EU) 2024/573 directly targets SF_6, NF_3, and CF_4—gases widely used in MEMS etching—signaling a definitive shift toward greener microfabrication practices.

Beyond regulatory mandates, many MEMS manufacturers are proactively aligning with global sustainability goals mentioned across the different government policies. Companies like STMicroelectronics, Bosch, and Infineon have set ambitious targets related to carbon neutrality, energy efficiency, and eco-friendly materials. STMicroelectronics aims to be carbon neutral by 2027, and Bosch's MEMS division in Reutlingen operates under ISO 14001 standards, emphasizing waste reduction and sustainable resource use.

In addition to corporate pledges, academic studies are reinforcing the case for sustainable design in fabrication environments. A 2021 study conducted by Vutla et al. presented an LCA of a MEMS cleanroom, revealing a high environmental sustainability score of 56.1 out of 60. The study quantified the cleanroom's energy efficiency using metrics like Net Total Pressure Efficiency (NTPE) and Total Energy Performance Index (TEPI), demonstrating the feasibility of green MEMS fabrication through architectural and operational improvements [14].

Together, these corporate and research efforts underscore a sector-wide shift toward environmental accountability and low-emission production strategies. The transition to green MEMS manufacturing is not only driven by policy and ethics but also by clear economic and market incentives. Reducing energy consumption and hazardous waste lowers operational costs, particularly in energy-intensive processes like plasma etching and vacuum deposition. As green technologies mature, fabs are finding that eco-efficiency directly translates to cost-efficiency.

Furthermore, integrating LCA into the design phase has proven instrumental in identifying environmental impacts and cost-saving opportunities. For instance, a study on MEMS piezoresistive pressure sensors [2] demonstrated that LCA-driven design can lead to more sustainable and economically viable products.

Recent research also confirms that environmental attitudes and digital green marketing significantly influence consumers' purchasing behavior and future buying intentions for sustainable products. A structural equation model by Štofejová et al. [15] demonstrated that willingness to pay for green products, environmental attitudes, and subjective norms all significantly shape consumer behavior—making green branding a compelling market strategy.

A notable example is the KISSES (Keep It Short, Simple, and Environmentally Sustainable) process developed by Lemaire et al. [16], which enables the fabrication of MEMS using recyclable materials such as aluminum foil, tree resin, and paper—without the need for cleanrooms. This low-energy, rapid prototyping approach demonstrates the feasibility of environmentally sustainable MEMS at low cost.

In parallel, novel academic efforts are also paving the way for environmentally conscious fabrication. Yang et al. (2022) introduced Eco resorbable and bioresorbable MEMS (eb-MEMS) that fully disintegrate into non-toxic end products, eliminating electronic waste. These devices are built from water-soluble materials using low-impact methods like stencil patterning and transfer printing—bypassing traditional cleanroom processes entirely [17].

On a broader scale, initiatives like SEMI's Sustainability Initiative and collaborative EU projects (e.g., Horizon Europe) are funding research into fluorine-free etching chemistries, AI-based process optimization, and material substitution strategies. These joint efforts aim to standardize greener alternatives and accelerate their industrial adoption, signaling a collective push toward sustainability that goes beyond isolated company efforts.

While the shift toward sustainable MEMS manufacturing is gaining momentum, several barriers still hinder large-scale adoption. The high capital cost of transitioning to fluorine-free or low-energy etching equipment remains a major constraint, especially for smaller fabs. Process compatibility and standardization also pose challenges, as many green alternatives are still in the experimental or pilot stage [18]. Additionally, the lack of unified sustainability benchmarks across the industry makes it difficult to measure and compare progress effectively.

However, with increasing regulatory pressure, corporate ESG commitments, and ongoing research breakthroughs, the outlook is promising. Overcoming these transitional hurdles will require cross-sector collaboration, policy incentives, and continued investment in scalable, clean fabrication technologies.

1.3 KEY GREEN ETCHING METHODS AND ALTERNATIVE APPROACHES

The transition to green etching technologies in MEMS fabrication is driven by the urgent need to minimize environmental and health hazards associated with traditional plasma and wet etching techniques. Conventional processes often rely on toxic fluorinated gases, hazardous acids, and energy-intensive vacuum systems that contribute significantly to greenhouse gas emissions, chemical waste, and high operational costs. In contrast, green etching emphasizes the substitution of these materials and methods with environmentally benign alternatives that maintain technical performance while improving sustainability. This section provides a critical examination of key green etching strategies, including dry and wet chemistries, hybrid systems, and solvent-free approaches, focusing on their mechanisms, compatibility with MEMS processes, and environmental advantages.

Gas-phase selective etching, especially for MXene synthesis, is a novel fluorine-free dry technique that eliminates the use of HF and other liquid-phase reagents. In this process, the A-layer element (typically aluminum or similar metals) in the MAX phase is selectively removed using reactive gas species like HCl or Cl_2, often assisted by moderate heating. This method operates without solvents or post-synthesis washing steps, drastically reducing chemical waste and exposure risks. Furthermore, it allows direct recovery of high-purity 2D MXene powders without generating fluoride-laden effluents, aligning strongly with the principles of green chemistry [19]. The process is scalable, energy-efficient, and suitable for batch processing, making it attractive for environmentally conscious nanomaterial production.

Chlorine-based dry etching, a mature alternative to fluorocarbon plasma etching, is particularly effective for high-aspect-ratio MEMS structures. Cl_2 plasmas generated in ICP or RIE systems form volatile by-products such as $SiCl_4$ or $AlCl_3$, which can be efficiently evacuated. One major advantage is the operation at lower pressures, which improves ion mean free path and enhances anisotropy by allowing reactive species to reach deeper features without excessive lateral etching. For instance, Cl_2-based etching of silicon can achieve aspect ratios up to 46:1 with vertical sidewalls, even in the absence of polymer-based sidewall passivation [20]. Unlike SF_6 or C_4F_8 plasmas used in the Bosch process, Cl_2 does not form carbon-based residues, thus reducing post-etch cleaning and associated solvent usage. This also contributes to longer equipment lifetimes and lower maintenance costs due to reduced polymer buildup in vacuum lines.

Bromine-based etch chemistries, including mixtures of HBr and Br_2 in HF or organic acid media, represent an emerging class of green wet etching systems. These mixtures offer comparable or superior etch rates to conventional HF–HNO_3 solutions while avoiding the generation of NOx gases, which are regulated air pollutants. Bromine's lower volatility reduces operator exposure and enhances surface smoothness, which is particularly advantageous for optical MEMS or solar applications where surface texturing is used to modulate reflectivity. The temperature-dependent etch rate of bromine-based solutions also enables tunable processing without aggressive heating, lowering the energy footprint of the etch process [21].

Green wet etching strategies aim to replace traditional mineral acids and solvents with biodegradable, less toxic compounds that maintain precision and material selectivity. Organic acids such as stearic acid and oleic acid have demonstrated functionality as surface-blocking agents, forming self-assembled monolayers (SAMs) on targeted surfaces to inhibit unwanted lateral etching. These acids are biodegradable, hydrophobic, and selectively adhere to hydrophilic or hydroxylated surfaces, enabling area-selective etching in microfabrication. Solvents like ethyl lactate and ethanol serve as green replacements for chlorinated or aromatic solvents used in vapor-phase processing or photoresist development. Moreover, ionic liquids (ILs) offer a tunable platform for etching and surface modification, with negligible vapor pressure, wide electrochemical windows, and high thermal stability. Their ability to be recycled and reused with minimal loss makes ILs particularly suitable for environmentally sensitive MEMS cleanroom applications [1].

Among hybrid etching approaches, supercritical CO_2 ($scCO_2$) processes have shown tremendous potential for reducing etching waste and avoiding structural damage. $ScCO_2$ possesses unique physicochemical properties such as low surface tension, high diffusivity, and gas-like viscosity, allowing it to penetrate nanoscale features without inducing capillary collapse—a common issue in aqueous drying steps. When combined with alcohol co-solvents and fluoride-based etchants, $scCO_2$ can etch sacrificial silicon dioxide layers with high rates (up to 500 nm/min) while preventing stiction, an essential requirement for suspended MEMS structures. This solvent-recycling, low-waste approach minimizes rinsing and drying stages and improves yield for complex geometries [22]. In addition, $scCO_2$ equipment often incorporates closed-loop systems for pressure and composition control, enabling consistent batch processing and reuse of CO_2, which enhances overall process sustainability.

Another hybrid technique that contributes to environmentally responsible etching is plasma-enhanced atomic layer deposition (PEALD) coupled with in-situ surface activation. While primarily a deposition method, this approach improves etch selectivity and reduces process steps through enhanced interface engineering. For example, TaCN films grown using Pentakis(dimethylamino)tantalum(V) Ta(N(CH$_3$)$_2$)$_5$ (PDMAT) precursors and H$_2$/Ar plasma demonstrate improved nucleation on oxide and low-k dielectrics after plasma pre-treatment. The plasma modifies the surface chemistry, increasing reactivity by creating dangling bonds or activating hydroxyl groups. This results in a gradual transition in bonding structure—from Ta–O to Ta–C and Ta–N—which reflects a shift toward more conductive and stable films. Because PEALD operates at relatively low temperatures and under vacuum, it generates minimal effluent and avoids the use of organic solvents, making it compatible with green fabrication strategies [23].

MXene gas-phase etching methods further enhance sustainability by operating under mild thermal conditions and allowing recovery of the volatile by-products without aqueous processing. Mechanistically, the selective removal of the A-layer is driven by the thermodynamics of volatile halide formation, which can be precisely controlled by adjusting temperature, partial pressure, and exposure time. These parameters enable uniform delamination, minimal surface contamination, and process scalability. Moreover, gas-phase approaches avoid challenges such as excessive washing, filtration, and drying encountered in conventional HF-based methods [19].

Table 1.1 provides a comparative summary of the green etching methods discussed, highlighting their technical features, process classifications, and specific environmental benefits.

Together, these emerging etching technologies represent a paradigm shift in MEMS manufacturing—one that embraces process integration, materials innovation, and environmental stewardship. However, these techniques are not without limitations. Issues such as the cost of precursor chemicals (e.g., ILs or specialized

TABLE 1.1

Summary of Green Etching Techniques and Their Environmental Benefits

Etching Method	Type	Key Features	Environmental Benefit
Chlorine-based etching (Cl_2)	Fluorine-free dry	Deep trench etching, vertical profiles, no polymer formation	Eliminates fluorinated gases; reduces post-etch residues
Bromine-based etching (HBr–Br_2)	Wet	High ambient temperature etch rate; reduced surface roughness	Avoids nitric acid; eliminates NOx and lowers chemical volatility
Organic acids & ionic liquids	Wet	Self-assembled monolayers; solvent-free resist stripping	Biodegradable; recyclable; reduced VOC emissions
Supercritical CO_2 (scCO_2)	Wet/hybrid	Deep penetration; anti-stiction; low surface energy	No liquid HF; fewer rinse cycles; closed-loop solvent recovery
Hybrid PEALD + surface activation	Hybrid	Plasma-assisted nucleation; low-temp processing	No solvent waste; improved layer conformity
MXene gas-phase etching	Dry	Solvent-free; direct powder recovery; scalable	Fluorine-free; no aqueous discharge; low energy input

plasmas), incompatibility with legacy equipment, and limited data on long-term stability hinder widespread adoption. Nonetheless, ongoing research into AI-optimized etch control, real-time endpoint detection, and closed-loop materials recovery offers promising pathways to overcome these barriers.

In conclusion, the integration of chlorine- and bromine-based plasma chemistries, organic and ionic wet etchants, hybrid processes using scCO_2 or plasma activation, and gas-phase routes for layered materials provides a rich toolkit for green MEMS etching. These methods reduce environmental burdens without compromising process fidelity, and with further standardization and industrial validation, they can become integral components of a sustainable microfabrication ecosystem.

1.4 CHALLENGES AND FUTURE PROSPECTS OF GREEN MEMS ETCHING

MEMS have significantly transformed various technological sectors by enabling miniature, high-performance devices. However, the environmental challenges posed by conventional MEMS fabrication techniques, particularly etching, necessitate the adoption of sustainable alternatives. Despite their promising benefits, green MEMS etching techniques encounter multiple barriers, including economic feasibility, scalability constraints, technological limitations, and complex process optimization requirements.

Economic viability is a significant hurdle in adopting green MEMS etching. Transitioning from conventional processes involves substantial upfront investments in new infrastructure and equipment. Although methods like directed self-assembly (DSA) and area selective deposition (ASD) theoretically reduce costs by eliminating photomasks and streamlining processes, their financial benefits remain speculative due to insufficient validation at industrial scales [1]. Additionally, the entrenched infrastructure for traditional dry etching, including costly vacuum chambers, gas handling systems, and safety measures, further complicates the economic feasibility of switching to green alternatives [24].

Scalability remains a critical issue in translating laboratory-level successes of green etching technologies to industrial-scale production. Techniques such as DSA and ASD face significant scalability hurdles, including challenges in maintaining process uniformity, defect control, and reproducibility, all essential for high-volume manufacturing [1]. Maintaining process stability across extensive wafer batches is technically demanding and has not yet been reliably achieved by green methods [24].

Green MEMS etching technologies are limited by technological and process-specific challenges. Fluorine-free dry etching methods employing chlorine and bromine plasmas are proposed as sustainable alternatives but face limitations such as tool corrosion, variable etch rates, and low selectivity, severely restricting their practical adoption [24]. Advanced plasma etching methods like neutral beam and radical-only plasmas show potential environmental benefits but suffer from process instability and limited industrial maturity [25].

Wet etching techniques, though cost-effective and simpler to implement, exhibit significant drawbacks including isotropy, reduced patterning precision, and hazardous chemical waste production, which diminishes their appeal as genuinely sustainable methods [24]. Hybrid etching approaches, which combine wet and dry etching processes, introduce additional complexities. Challenges include precise alignment and controlling multi-step procedures, complicating their integration into mainstream manufacturing [24]. Hybrid methods involving Atomic Layer Etching (ALE) and ASD often experience low throughput and stringent process control requirements, further hindering scalability [25].

Recent developments in molten salt Lewis acidic etching provide promising solutions to traditional fluorine-based etching's environmental and safety concerns. This innovative method is not only fluorine-free but also capable of precise, environmentally safer etching, with the flexibility to tailor surface chemistry [26]. Hybrid approaches involving molten salt etching and subsequent functionalization steps enhance material functionality and present innovative integration possibilities for sustainable MEMS fabrication [26].

Artificial intelligence (AI) plays a transformative role in optimizing green MEMS etching processes, significantly enhancing their feasibility. AI-driven models and digital twins can predict outcomes, optimize plasma conditions, and minimize waste, thereby increasing process efficiency and reducing environmental impact [25]. An attention-based Long Short-Term Memory (LSTM) machine learning model achieved remarkable accuracy (98.2%) in plasma etch endpoint detection using Optical Emission Spectroscopy (OES) data, highlighting the potential of AI to reduce material usage and enhance process reliability [27]. Similarly, deep reinforcement learning (Deep Q-Networks) has been effectively used to optimize parameters for DRIE, achieving precise etch profiles while minimizing manual interventions, resource waste, and energy consumption [28].

Several emerging technologies indicate promising directions for future sustainable MEMS etching. Atmospheric downstream plasma etching stands out as a low-temperature, environmentally friendly technique capable of significantly reducing the environmental footprint [24]. Innovative methods such as cryogenic etching, halogen-free plasma chemistries, and sustainability-oriented equipment designs offer promising avenues for future green MEMS fabrication [25]. Additionally, molten salt Lewis's acid etching shows substantial potential for scale-up and commercial implementation due to its effectiveness and broader material compatibility, indicating a practical pathway toward future green manufacturing [26].

In summary, while green MEMS etching presents considerable environmental and operational benefits, substantial challenges persist. Economic uncertainty, scalability limitations, technical immaturity, and complex integration barriers must be addressed through continued innovation. Advancements in hybrid methods, AI-driven process optimization, and new material chemistries, coupled with collaboration among academia, industry, and regulatory bodies, are critical for the widespread adoption and success of green MEMS etching technologies.

REFERENCES

1. Mullen, E., & Morris, M. A. (2021). Green nanofabrication opportunities in the semiconductor industry: A life cycle perspective. *Nanomaterials*, 11(5), 1085.
2. Delhaye, T. P., Brun, G. L., Flandre, D., & Raskin, J. P. (2021). Bottom-up life-cycle assessment of MEMS piezoresistive pressure sensors. In *2021 Symposium on Design, Test, Integration & Packaging of MEMS and MOEMS (DTIP)* (pp. 1–6). IEEE.
3. Intergovernmental Panel on Climate Change (IPCC). (2001). *Good practice guidance and uncertainty management in national greenhouse gas inventories, Chapter 3: Industrial processes sector—PFC, HFC, NF_3 and SF_6 emissions from semiconductor manufacturing*. IPCC.
4. Prather, M. J., & Hsu, J. (2008). NF_3, the greenhouse gas missing from Kyoto. *Geophysical Research Letters*, 35, L12810.
5. Weiss, R. F., Mühle, J., Salameh, P. K., & Harth, C. M. (2008). Nitrogen trifluoride in the global atmosphere. *Geophysical Research Letters*, 35, L20821.
6. Nagano, S., Shibata, T., Sakoda, K., Inoue, M., Hasaka, S., Takano, T., Ikehara, T., & Maeda, R. (2007). Environment friendly MEMS fabrication: Proposal of new D-RIE process gases for reduction of greenhouse effect. In *Proceedings of IEEE 20th International Conference on Micro Electro Mechanical Systems (MEMS)* (pp. 341–344). IEEE.
7. Shen, C. W., Tran, P. P., & Minh Ly, P. T. (2018). Chemical waste management in the U.S. Semiconductor industry. *Sustainability*, 10(5), 1545.
8. Ruberti, M. (2024). Environmental performance and trends of the world's semiconductor foundry industry. *Journal of Industrial Ecology*, 28, 1183–1197.
9. Wang, Q., Huang, N., Chen, Z., Chen, X., Cai, H., & Wu, Y. (2023). Environmental data and facts in the semiconductor manufacturing industry: An unexpected high water and energy consumption situation. *Water Cycle*, 4, 47–54.
10. Boyd, S. B., Horvath, A., & Dornfeld, D. (2009). Life-cycle energy demand and global warming potential of computational logic. *Environmental Science & Technology*, 43(19), 7303–7309.
11. Oehrlein, G. S., Brandstadter, S. M., Bruce, R. L., Chang, J. P., DeMott, J. C., Donnelly, V. M., Dussart, R., Fischer, A., Gottscho, R. A., Hamaguchi, S., Honda, M., Hori, M., Ishikawa, K., Jaloviar, S. G., Kanarik, K. J., Karahashi, K., Ko, A., Kothari, H., Kuboi, N., Kushner, M. J., Lill, T., Luan, P., Mesbah, A., Miller, E., Nath, S., Ohya, Y., Omura, M., Park, C., Poulose, J., Rauf, S., Sekine, M., Smith, T. G., Stafford, N., Standaert, T., & Ventzek, P. L. G. (2024). Future of plasma etching for microelectronics: Challenges and opportunities. *Journal of Vacuum Science & Technology B*, 42(4), 041501.

12. Ruberti, M. (2023). The chip manufacturing industry: Environmental impacts and eco-efficiency analysis. *Science of The Total Environment*, 858(Part 2), 159873.
13. IRDS Environmental Safety and Health and Sustainability (ESHS) and Equipment, Support, and Services Factory (ESSF) Team. (2024). *2024 IRDS ESHS-ESSF regulatory and commitments white paper.* International Roadmap for Devices and Systems. https://irds.ieee.org/images/files/pdf/2024/2024IRDS_ESHS-ESSF_Regulatory_and_Commitments_White_Paper.pdf
14. Vutla, S. R., Regalla, S. P., & Ramaswamy, K. (2021). Life cycle assessment of cleanroom for micro-electro-mechanical systems fabrication with insights on sustainability. *Journal of Cleaner Production*, 282, 124520.
15. Štofejová, L., Kráľ, Š, Fedorko, R., Bačík, R., & Tomášová, M. (2023). Sustainability and consumer behavior in electronic commerce. *Sustainability*, 15(22), 15902.
16. Lemaire, E., Thuau, D., Caillard, B., & Dufour, I. (2015). Fast-fabrication process for low environmental impact microsystems. *Journal of Cleaner Production*, 108, 207–216.
17. Yang, Q., Liu, T. L., Xue, Y., et al. (2022). Ecoresorbable and bioresorbable microelectromechanical systems. *Nature Electronics*, 5, 526–538.
18. Semiconductor PFAS Consortium. (2023). *Plasma etch and deposition white paper.* Semiconductor Industry Association. https://www.semiconductors.org/wp-content/uploads/2023/06/FINAL-Plasma-Etch-and-Deposition-White-Paper.pdf
19. Zhu, J., Zhu, S., Cui, Z., et al. (2024). Solvent-free one-step green synthesis of MXenes by "gas-phase selective etching". *Energy Storage Materials*, 70, 103503.
20. Tian, W. C., Weigold, J. W., & Pang, S. W. (2000). Comparison of Cl_2 and F-based dry etching for high aspect ratio Si microstructures etched with an inductively coupled plasma source. *Journal of Vacuum Science & Technology B*, 18(4), 1890–1896.
21. Thøgersen, A., Andreasen, J. S., Andersen, K. B., & Petersen, N. R. (2024). Comparison of acidic and alkaline silicon etching solutions using industrial alkaline textured silicon for solar cells as a reference. *Green Chemical Engineering*, 3, 100104.
22. Saga, K., & Hattori, T. (2007). Wafer cleaning using supercritical CO_2 in semiconductor and nanoelectronic device fabrication. *Solid State Phenomena*, 134, 97–102.
23. Reif, J., Knaut, M., Killge, S., Albert, M., & Bartha, J. W. (2019). In vacuo investigations on the nucleation of TaCN by plasma enhanced atomic layer deposition. *Microelectronic Engineering*, 211, 13–17.
24. Madou, M. J. (2018). *Fundamentals of microfabrication and nanotechnology* (3rd ed.). CRC Press.
25. Kanarik, K. J., et al. (2024). Future of plasma etching for microelectronics: Challenges and opportunities. *Journal of Vacuum Science & Technology B*, 42(4), 041501.
26. Siddique, S., Liu, X., Zhang, Y., & Wang, P. (2023). Fluorine-free MXenes via molten salt Lewis acidic etching: Applications, challenges, and future outlook. *Progress in Materials Science*, 139, 101183.
27. Kim, Y. J., et al. (2024). Improved plasma etch endpoint detection using attention-based long short-term memory ML. *Electronics*, 13(17), 3577. https://doi.org/10.3390/electronics13173577
28. Wang, F., et al. (2025). Deep reinforcement learning-based parameters optimize prediction model for smooth-vertical sidewall profile in DRIE. In *Proceedings of 2025 IEEE MEMS* (pp. 565–568).

2 Theoretical Foundations and Mechanisms of Etching

2.1 FUNDAMENTALS OF MATERIAL REMOVAL IN ETCHING

Etching stands as an essential fabrication technique in manufacturing microelectro-mechanical systems (MEMS), and their biological counterparts, bioMEMS. Etching is indispensable in MEMS for creating intricate 3D structures and high-aspect-ratio features necessary for precise control over device geometry and functionality. Etching allows for the selective removal of sacrificial layers to form suspended components, ensuring mechanical stability and enabling motion or sensing capabilities critical to MEMS performance. Etching is fundamentally a subtractive technique in microma-chining of microsystems, as its goal is to remove material from the substrate, or other material layers in a localized and controlled manner to achieve a pattern transfer. Etching techniques are typically categorized along four aspects: chemical, electro-chemical, thermal, or mechanical processes, wet versus dry methods, and isotropic or anisotropic outcomes, depending on the mechanism and directionality of mate-rial removal. However, some etching processes can combine different techniques to achieve optimized pattern transfer, for example, chemical-physical processes. These categorizations arise from fundamental principles such as atomic-scale interactions like adsorption, desorption, and surface reactions, reaction kinetics governing etch rates and uniformity, and the selectivity of the etching process. The etching process can be described and controlled through etching parameters that include etch rate, uniformity, throughput, directional control, and selectivity of etch. The following sections will clarify the distinctions between chemical and physical etching, along-side other critical factors that define etching processes, such as selectivity, direction-ality, and the interplay between surface reactions and energy transfer [1, 2].

Chemical etching and physical etching represent two distinct modes of material removal, which are differentiated by their underlying mechanisms. Chemical etch-ing is driven by chemical reactivity, whereas physical etching is driven by kinetic energy transfer [1, 2]. In chemical etching, the removal of material driven by chemi-cal reactivity is also referred to as reaction-driven chemical etching. In this mode, the etchant (which can be either a liquid solution or a reactive gas) starts a chemical reaction with the substrate or other material layer, breaking atomic bonds to form volatile products that desorb from the surface. Desorption refers to atoms releasing from a surface into the surrounding environment. Chemical etching is reliant on the thermodynamic favorability of the reaction and the activation energy required to overcome bonding forces. Etching selectivity, an etching parameter, comes about from the etchant's specificity. In other words, the etchant reacts preferentially with

DOI: 10.1201/9781003674795-2

TABLE 2.1

Comparison of Mechanism, Selectivity, Directionality between Wet and Dry Etching [1–8]

Aspect	Wet Etching	Dry Etching
Mechanism	Chemical dissolution in liquid etchants.	Chemical and/or physical reactions (plasma or ions).
Selectivity	High (material-specific).	Moderate (chemical) / Low (physical).
Directionality	Isotropic (non-directional).	Anisotropic (directional, typically vertical).

certain materials while leaving others intact or less affected. Other etching parameters like etch rate and uniformity are linked to surface phenomena such as adsorption of reactive species and desorption of chemical byproducts [3–6]. On the other hand, physical etching is energy-driven in the sense that it requires high-energy particles that collide with the substrate surface or other material layer resulting in the physical dislodgement of atoms through the transfer of momentum. In this mode of etching, no chemical reactions occur and material removal results entirely from the kinetic energy of incoming particles. Etching parameters important in physical etching include ion energy, flux, and incident angle of the colliding particles. These parameters directly influence the etch rate and feature geometry. It is reasonable to conclude that this method is less selective as any material subject to sufficiently energetic collisions will erode. It, nonetheless, offers precise control over directionality, which permits highly anisotropic profiles for high-aspect-ratio microstructures [1, 2, 7, 8].

Table 2.1 shows comparison table summarizing key differences between wet and dry etching. (Source: Self-made, [1–8]). Notes: Wet etching is cost-effective but limited by isotropy, while dry etching enables precise, high-aspect-ratio 3D structures. Chemical dry etching (e.g., reactive ion etching) combines selectivity and directionality, whereas physical methods (e.g., ion milling) prioritize directionality over material specificity.

Directionality in etching: Isotropic vs. anisotropic outcomes: In etching processes, the terms isotropy and anisotropy describe how material removal varies spatially (Figure 2.1). Isotropic etching occurs uniformly in all directions creating rounded or undercut profiles. This uniformity arises when diffusion or reaction rates dominate over directional influences. In contrast, anisotropic etching removes

ISOTROPIC ETCHING   **ANISOTROPIC ETCHING**

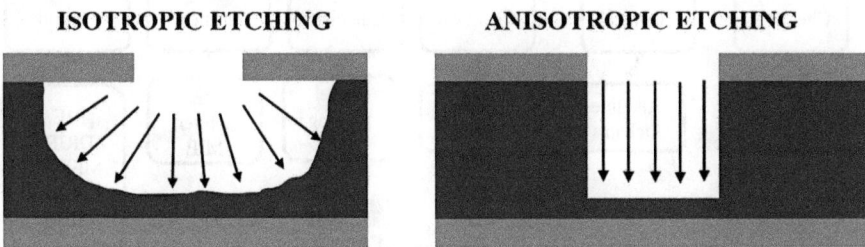

FIGURE 2.1   Isotropic versus anisotropic etching profiles. (Source: Self-made [1–8].)

material directionally, forming vertical sidewalls with minimal lateral erosion. This directionality can arise from energy-driven mechanisms, such as particle bombardment, where particles strike the surface at near vertical angles, transferring momentum along specific axes. While isotropy and anisotropy primarily describe spatial uniformity, they indirectly have an influence over etching selectivity. For example, anisotropic methods often achieve selectivity through directional energy delivery rather than chemical affinity, enabling patterning of materials that might otherwise resist isotropic etchants [1, 2].

Etching can also be classified by the nature of the etchant, as either dry or wet. Wet etching employs liquid-phase chemicals that can be acids, bases, or other solvent-based solutions [3–6]. Dry etching mostly requires the use of gaseous reactants or plasma to remove materials. However, dry etching processes often combine chemical reactions and physical interactions. Furthermore, there is the common misconception that dry etching is restricted to plasma techniques, but there exist methods of dry etching that do not require plasmas [7, 8]. Etching processes are broadly defined by their medium as either wet or dry, directionality as isotropic or anisotropic, and the underlying removal mechanisms, which may involve chemical reactions, physical energy transfer, electrochemical interactions, or hybrid approaches that integrate multiple principles.

These classifications are not mutually exclusive. For instance, reactive ion etching (RIE) combines chemical reactivity with ion bombardment, leveraging the directional control from physical processes with the selectivity of chemical etching. Figure 2.2 summarizes the primary classifications of etching techniques, organized by their medium as either wet or dry, directionality as isotropic or anisotropic, and underlying material removal mechanisms [1–8].

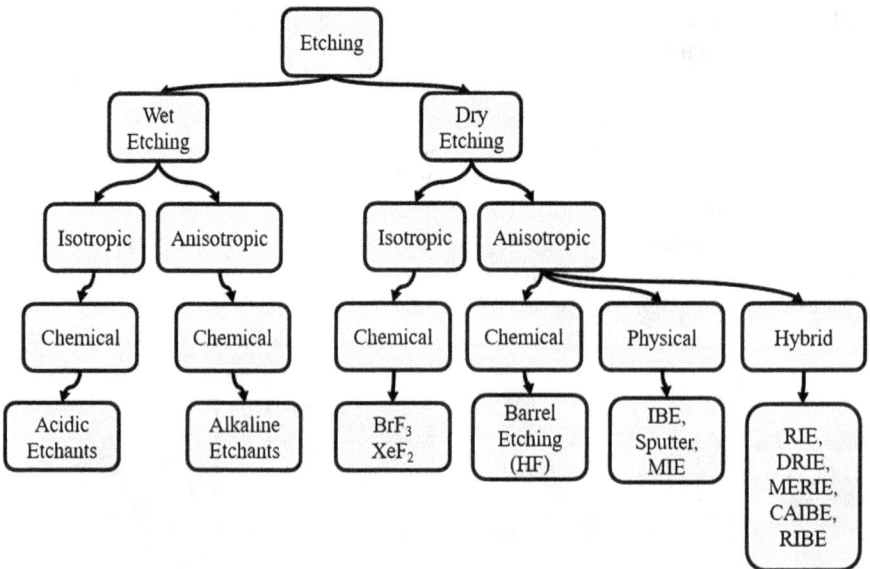

FIGURE 2.2 Classification of most common etching techniques. (Source: Self-made [1–8].)

As seen from Figure 2.2, wet etching is mostly a chemically-driven technique that can be either isotropic or anisotropic. It can be immediately noted that isotropic wet etching predominantly requires of acidic etchants. This is due to several reasons like the solubility of reaction byproducts, proton-driven bond weakening, selective reactivity, inhibition of passivation, compatibility with isotropic kinetics, and practical industrial considerations. Firstly, acidic etchants are utilized for isotropic wet etching due to the fact that acidic etchants promote the formation of water-soluble products that can be easily and continuously removed without any resulting surface passivation. For example, etching silicon dioxide (SiO_2) with hydrofluoric acid (HF) results in the formation of stable hexafluorosilicic acid (H_2SiF_6) along with water in an aqueous solution that can be easily removed, while nitric acid (HNO_3) oxidizes silicon to soluble SiO_2. Furthermore, acids donate protons or H+ ions, which protonate oxygen atoms in oxide materials like silicon oxide and aluminum oxide, weakening metal-oxygen bonds. This metal-oxygen bond weakening lowers the activation energy (E_a), which enables dissolution and thus faster and uniform etching. For example, HF reacts with SiO_2 by protonating its oxygen atoms and breaking its Si-O bonds to form volatile silicon tetrafluoride (SiF_4) and water or other fluorosilicates that are soluble. Another reason for the employment of acidic etchants in isotropic wet etching comes from acidic etchants' selective reactivity, which enables us to select specific etchants to target specific materials. Acid etchants are also favored for isotropic wet etching because of their property of inhibiting passivation. Many metals like copper and silver dissolve via oxidation in acid etchants without forming passivating oxide layer, in contrast to alkaline etchants that often form insoluble hydroxides that passivate surfaces and thus limit isotropic removal. Moreover, acid etchants are compatible with isotropic kinetics, where diffusion-limited or reaction-limited kinetics favor uniform material removal. In diffusion-limited cases, stirring enhances mass transport, while temperature adjustments enhance reaction-limited processes. Finally, practical industrial considerations favor the use of acid etchants for isotropic wet etching. Acid concentration and temperature are easily adjusted to control etch rates and selectivity, as can be seen from iso-etch curves for isotropic wet etching [4–6].

Many etching parameters like selectivity, etch rate, uniformity, etc. have already been mentioned but not thoroughly discussed. These parameters are going to be thoroughly introduced in the context isotropic wet etching. Etch rates, or the rate at which a material or film is removed by an etchant, are determined by several phenomena of the reaction. The etch rate is primarily influenced by the slowest step in the etching reaction sequence. The etching reaction sequence can be summarized as follows: first, the etchant is transported to the surface through diffusion into a boundary layer followed by its adsorption of the reactive species onto the substrate from where the chemical reaction starts, resulting in bond-breaking and byproduct formation and desorption of the soluble byproducts and the mass transport of byproducts away from the surface [4–6, 9]. In reaction-limited isotropic wet etching processes, the etching rate depends on the energy required to break bonds as dictated by the Arrhenius model of reaction rates:

$$k = Ae^{\frac{-E_a}{RT}} \qquad (2.1)$$

where k, or the reaction rate, increases exponentially with temperature (T) when the activation energy (E_a) remains constant [10]. In diffusion-limited isotropic wet etching processes, mass transport dictates etching rates, where stirring reduces the boundary layer thickness and thus enhances the delivery of etchant and byproduct removal. This can be modeled with Fick's laws of diffusion that explain how etchant viscosity and concentration gradients influence etching rates [11].

Now, etch-selectivity or the relative etch rates between mask, films, substrate, or other materials, can be explained from the differential chemical reactivity between materials. For example, HF etches SiO_2 rapidly while leaving silicon (Si) mostly intact due to SiO_2's lower activation energy and favorable Gibbs free energy change (ΔG) for dissolution. Selectivity, however, can be modulated through etchant concentration, pH, and the use of additives like ammonium fluoride (NH_4F). Etching uniformity is dependent on the homogeneous adsorption and reaction across the substrate [2, 4–6, 9, 10, 12]. The Langmuir adsorption model is useful in describing the adsorption of etchant molecules onto the surface in wet etching. Affinity and etchant concentration have a direct influence on surface coverage (Θ):

$$\theta = \frac{KC}{1+KC} \tag{2.2}$$

where K is the adsorption equilibrium constant and C is the etchant concentration. Higher surface coverage increases the reaction probability, but the reaction may saturate at high etchant concentrations. The desorption of reaction products is also important as incomplete desorption can blockade active surface sites, impeding further reaction [12]. In some cases, surfactants like polyethylene glycol (PEG) can be utilized to modify the surface tension and facilitate byproduct mass transfer, enhancing reaction uniformity [13]. In etching rates and selectivity, thermodynamics is responsible for reaction feasibility, while kinetics or activation energy dictates the speed of the reaction. A useful equation to model chemical kinetics in wet etching is the Eyring equation, which describes changes in chemical reaction rate (e.g., etch rate) in response to temperature:

$$k = \frac{\kappa k_B T}{h} e^{-\frac{\Delta G}{RT}} \tag{2.3}$$

where k is the reaction rate, k_B is Boltzmann's constant, h is Planck's constant, T is the temperature, ΔG is the Gibbs energy of activation, and κ is the transmission coefficient [10, 14]. When performing isotropic wet etching, it is useful to consult iso-etch curves, which map etch rates as functions of temperature and etchant concentration. These empirical curves aid in etch optimization and tuning for etch rate, selectivity, and uniformity [5, 6, 9]. In anisotropic wet etching of silicon, etch rate, selectivity, and uniformity are crystal orientation dependent and require the use of alkaline etchants. Alkaline etchants are superior in their ability to selectively attack silicon crystal planes depending on differences in atomic bonding densities across different crystal planes [2–6, 9]. The following sections discuss in detail chemical and plasma etching mechanisms, electrochemical and hybrid etching principles, and factors influencing etching efficiency and surface quality, all of which are indispensable and fundamental in the microfabrication process of MEMS and bioMEMS devices.

2.2   CHEMICAL AND PLASMA ETCHING MECHANISMS

Etching processes are fundamental to the fabrication of MEMS, enabling the controlled removal of specific material layers to create functional structures with high precision and resolution. These processes are typically categorized into chemical (wet) and plasma-based (dry) etching mechanisms. Each category is governed by distinct physical principles and offers different capabilities with respect to anisotropy, selectivity, material compatibility, throughput, and environmental impact. As sustainable manufacturing gains increasing emphasis, understanding the detailed mechanisms behind each etching approach becomes essential for the development and integration of fluorine-free alternatives in green MEMS fabrication.

Wet chemical etching, one of the oldest and most widely used methods, involves immersing the substrate in a liquid-phase chemical solution that reacts with the target material to form soluble byproducts. The etching profile is often isotropic due to uniform diffusion of reactive species in all directions, although anisotropic wet etching is achievable under specific conditions, especially with crystalline substrates like silicon. The rate of material removal depends on multiple interrelated parameters, including the etchant composition and concentration, temperature, solution agitation, and the doping level of the substrate. For example, doped silicon typically exhibits higher etch rates compared to intrinsic silicon due to the increased density of charge carriers, which facilitates bond breaking during chemical attack [15, 16].

Two primary reaction mechanisms dominate wet chemical etching: acid-base dissolution and redox-driven electrochemical reactions. In acid-base etching, strong acids or bases engage in chemical reactions that directly dissolve the target material. A common example is the mixture of HNO_3 and HF, which is used to etch silicon. In this two-step mechanism, HNO_3 oxidizes elemental silicon to form SiO_2, which is subsequently dissolved by HF. The global reaction pathway includes the oxidation of silicon to SiO_2, followed by its conversion into H_2SiF_6, a highly soluble product. This combination of reactions allows for rapid and controllable etching. Despite its widespread use, HF presents significant health and environmental hazards due to its extreme toxicity and ability to penetrate biological tissues. Even dilute HF solutions are capable of etching SiO_2 effectively, making it both indispensable and dangerous in microfabrication environments [17, 18].

A more advanced and highly anisotropic form of wet etching is metal-assisted chemical etching (MacEtch), a redox-driven process that relies on a patterned metal layer such as gold or silver deposited on the silicon substrate. This metal acts as a catalyst for localized redox reactions, facilitating the injection of holes (h^+) into the silicon valence band, which enhances the oxidation of silicon directly beneath the metal. The oxidized silicon is then selectively dissolved in the presence of HF. Because the reaction is localized to the metal-covered regions and proceeds preferentially in the vertical direction, MacEtch offers exceptional anisotropy, allowing the formation of deep, narrow features with minimal undercut. It is especially effective for applications requiring high-aspect-ratio etching without the complexity of lithographic sidewall protection [19, 20].

In contrast to wet etching, plasma etching—or dry etching—relies on the interaction between ionized gas species and the substrate surface. The process typically occurs in a low-pressure vacuum chamber where a radio-frequency (RF) electromagnetic field excites process gases into a plasma state, generating a reactive mixture of ions, electrons, radicals, and neutral species. These energetic components interact with the surface through a variety of mechanisms that include physical sputtering, chemical reactions, and ion-enhanced etching. The combination of these mechanisms enables precise pattern transfer and vertical etch profiles, particularly important for high-resolution microfabrication.

Physical sputtering, also known as ion bombardment, involves the acceleration of positive ions, such as Ar^+, toward the substrate surface under the influence of the plasma sheath's electric field. Upon impact, the ions transfer kinetic energy to surface atoms, physically dislodging them from the material through momentum exchange. This process is highly directional and enables excellent anisotropy, making it suitable for creating well-defined vertical profiles. However, physical sputtering lacks material selectivity because the energy-based mechanism does not differentiate between chemical compositions. It can also lead to substrate damage, especially when high ion energies are used without adequate process control [21].

Chemical plasma etching, in contrast, relies primarily on reactive neutral species—particularly radicals—generated within the plasma. These radicals, such as fluorine (F•), chlorine (Cl•), and bromine (Br•), chemically interact with the substrate to form volatile products that are subsequently evacuated from the chamber. Because radicals are uncharged and have long mean free paths (MFPs), they diffuse uniformly and react isotropically, often leading to undercutting unless countermeasures are taken. Nonetheless, radical-based etching offers high selectivity due to the chemical specificity of radical-substrate interactions. This makes it advantageous for processes where selectivity over masking materials is essential, such as in the etching of SiO_2 in the presence of silicon [22, 23].

Ion-enhanced chemical etching, also referred to as RIE, combines the directionality of ion bombardment with the reactivity of radicals. In RIE, low-energy ions physically assist the removal of passivation layers or surface contaminants, while radicals chemically react with freshly exposed material to form volatile byproducts. This synergistic mechanism provides high anisotropy and improved etch rate compared to either sputtering or radical-based etching alone. RIE enables excellent profile control and is widely used in applications requiring submicron precision. However, because ions lose energy through collisions within the plasma sheath and process gases, fine-tuning ion energy and density is necessary to achieve optimal performance without compromising selectivity or causing damage to sensitive structures [24].

A fourth mechanism commonly integrated into advanced plasma etching processes is sidewall passivation. This technique is essential in deep reactive ion etching (DRIE), where polymer films are cyclically deposited on the sidewalls to inhibit lateral etching and preserve vertical profiles. In the Bosch process, for example, an SF_6 plasma is used to etch vertically, while a C_4F_8 plasma deposits a fluorocarbon polymer that passivates the sidewalls. Alternating between these steps enables the

| Ion Bombardment | Radical Reaction Etching | Reactive Ion Etching | Reactive Ion Etching with sidewall passivation |

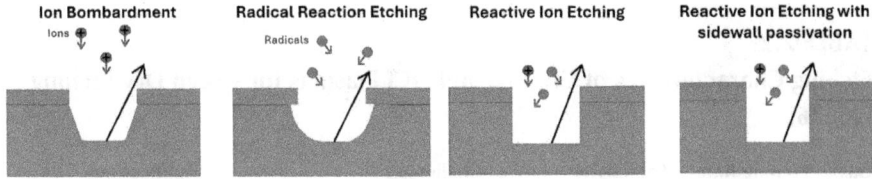

FIGURE 2.3   Schematic representation of four key plasma etching mechanisms: (1) Physical sputtering; (2) Radical-assisted etching; (3) Ion-enhanced chemical etching (RIE); (4) Etching with sidewall passivation.

formation of high-aspect-ratio trenches with scalloped sidewalls. The control of passivation thickness, deposition rate, and ion energy during the etch step determines the final profile geometry and uniformity. Sidewall passivation has become a cornerstone of MEMS fabrication due to its ability to deliver tall, narrow structures with exceptional reproducibility.

As illustrated in Figure 2.3, these four etching mechanisms—sputtering, radical chemistry, RIE, and sidewall passivation—differ markedly in their etch profiles and processing behavior. Each mechanism serves a unique role in defining etch directionality, selectivity, and rate, and often they are used in combination to meet complex fabrication requirements.

Historically, fluorinated gases such as CF_4, SF_6, and NF_3 have been the workhorses of plasma etching due to their high reactivity with silicon-based materials and their capacity to form volatile etch products. However, these gases also exhibit extremely high global warming potentials (GWP), with SF_6 and NF_3 among the most potent greenhouse gases known. Their environmental persistence and the challenges associated with abatement have prompted a shift toward fluorine-free alternatives such as Cl_2, Br_2, and HCl, which offer reduced ecological impact and improved sustainability [25].

Chlorine gas (Cl_2) provides high etch rates and sharp profile definition, making it suitable for patterning silicon and certain metals. Its high reactivity enables vertical profiles even without polymer-based passivation, although process optimization is needed to avoid micro-masking and residue formation. Bromine gas (Br_2) offers lower etch rates but better selectivity and smoother surface finishes, making it ideal for delicate MEMS structures and DRAM fabrication. Hydrogen chloride (HCl), although generally slower and less aggressive, is often used in conjunction with other gases for chamber cleaning, surface preparation, or controlled oxide removal. Each of these alternatives exhibits unique chemical properties that can be exploited for specific applications, with the trade-offs involving etch rate, material compatibility, and residue management [26, 27].

The comparative characteristics of these fluorine-free gases are summarized in Table 2.2, which outlines their relative etch performance and process advantages in typical MEMS applications.

In conclusion, etching mechanisms in MEMS fabrication encompass a rich array of physical and chemical interactions that can be tailored to achieve precise structural and functional outcomes. From the isotropic, solution-based dissolution of wet chemical etching to the anisotropic, plasma-assisted precision of RIE and DRIE,

TABLE 2.2
Etching Characteristics of Cl_2, Br_2, and HCl Plasmas for Green Dry Etching [25, 26]

Gas	Etching Rate	Selectivity	Challenges	Applications
Cl_2	High	Moderate	Requires passivation control; toxic gas handling	Precision silicon/metal patterning
Br_2	Medium	High	Slower rate; better for smooth delicate features	MEMS trenching; DRAM fabrication
HCl	Low	Moderate	Limited reactivity; often used for cleaning steps	Oxide removal; chamber conditioning

each mechanism offers unique advantages and limitations. As the industry moves toward greener and more sustainable practices, the adoption of fluorine-free etching chemistries and hybrid plasma processes represents a promising path forward. Understanding the technical nuances of each mechanism—combined with advances in in-situ diagnostics and process modeling—will be critical for developing next-generation etching platforms that meet both performance and environmental targets.

2.3 ELECTROCHEMICAL AND HYBRID ETCHING PRINCIPLES

Electrochemical etching is a precise wet etching method based on externally provided electrical potential to cause reactions at the surface of a conductive material that is in an electrolyte solution. It causes controlled oxidation and selective removal of material, often metal or semiconductor, at the anode. The material is etched and dissolves into the electrolyte as ions as a result of the reaction [28].

The etching process requires two electrodes: the working electrode and a counter electrode. On applying the potential difference, the working electrode acquires an anodic charge and induces localized dissolution by electrochemical reactions. Etching rate, pattern, and selectivity vary with the nature of the electrolyte, voltage, current density, and exposure time. Electrochemical etching has a lot of applications in micro- and nanofabrication for high-aspect-ratio structure generation, porous silicon, or smooth patterned surfaces. In the selective dissolution of silicon, the material surface is first oxidized (Si $\rightarrow$ SiO_2). The resulting oxidized layer is soluble in the electrolyte, with HF dissolving the SiO_2. Oxidation and subsequent dissolution occur only in the electrically activated areas.

KEY PARAMETERS

- Voltage: Controls the oxidation rate which means it also controls the overall etch rate, depth, and porosity of the structure.
- Current: If voltage is constant, current can be pulsed to have finer control over the porosity. Pulsed etching also helps in reducing bubble formation and facilitates self-termination during tip shaping that is discussed in applications section.

TABLE 2.3

Common Electrolytes and Related Materials for Electrochemical Etching

Material	Common Electrolyte
Si	HF-based (e.g., HF + ethanol)
W	NaOH, KOH, HNO$_3$, or HF
Metals	Acids like HCl, HNO$_3$, H$_2$SO$_4$

- Electrolytes: Table 2.3 lists the comparison of most common electrolytes for etching some related materials (Si, W, metals, etc.).

In addition, HF is used in porous Si because it dissolves SiO$_2$, allowing etch to continue. Moreover, Ethanol is often added to reduce surface tension and improve wetting, preventing bubble interference.

Gas bubbles can be formed on the electrode surface during electrochemical etching. Bubbles will adhere to the surface and create etching voids, inhomogeneities, or localized etch stop regions if not controlled [28], which is highly unwanted in high-precision microtool fabrication. Mitigation of that bubbling effect can be done by means of electrolyte management. Adding solvents like ethanol, mechanical stirring of the electrolyte, pulsed current etching—which is discussed in the previous section—and active flowing electrolyte that refreshes the electrolyte constantly are all forms of electrolyte management that reduce the bubble formation [29–31]. Electrolyte management is also required in order to maintain uniform chemical activity. Over time, electrolyte composition changes as dissolved material accumulates, and reactive species are consumed, which reduces etch uniformity and increases defect levels.

Electrochemical etching can be used for the microfabrication of microtools such as ultra-sharp profilometry tips and Atomic Force Microscopy tips [32, 33]. It can be achieved by immersing the electrode in electrolyte for a specific amount of time till it reaches the desired shape and diameter. After the electrode reaches the desired diameter, the position controller retracts it for the subsequent process. Then, the immersed part of the electrode is etched again. This procedure finally produces a stepped shape. Conical shape electrodes can be produced by pulling out the microtool continuously in very short time steps as seen in Figure 2.4.

As discussed in Section 2.1, dry etching is separable into physical etching, chemical etching, or both. Ion beam etching (IBE)—also referred to as ion etching or ion-beam milling—is one of the purely physical etching methods. In IBE, a high-energy inert ion beam, typically argon (Ar$^+$), is accelerated onto the target substrate under vacuum. The incident ions transfer their kinetic energy directly into the surface atoms of the material, physically pushing them away without chemical reactions. This allows for highly anisotropic etching profiles and precise material removal. Because IBE relies solely upon momentum transfer, it has excellent etch depth and uniformity control but will etch more slowly than chemical-assisted techniques [1].

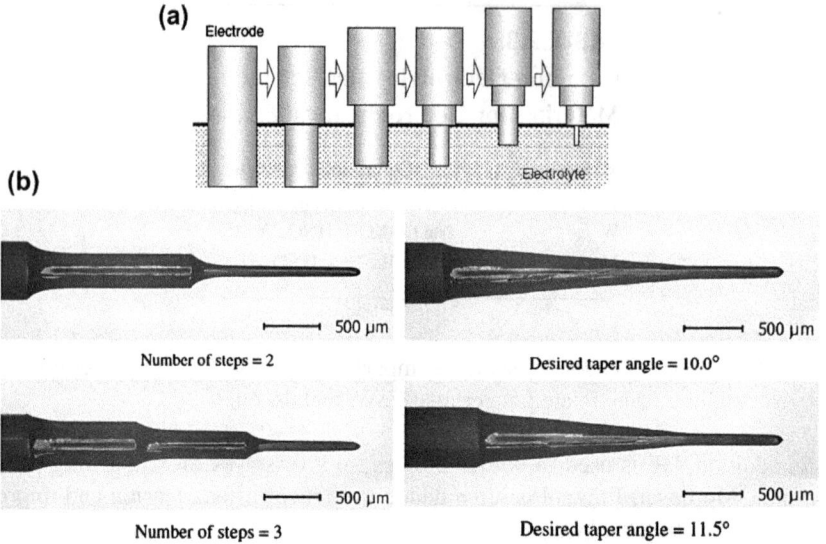

FIGURE 2.4 (a) Conceptual drawing of the fabrication of a stepped shape microtool, (b) actual fabricated microtools with different steps and taper angles by etching [33].

Figure 2.5 consists of an ICP and parallel carbon plates. The process chamber is separated from the plasma chamber by a carbon plate fitted at the bottom, because carbon has the lowest sputtering yield under high-energy bombardment and does not contaminate semiconductor devices. There are numerous apertures in the bottom carbon plate to extract neutral beams from the plasma in the process chamber [34].

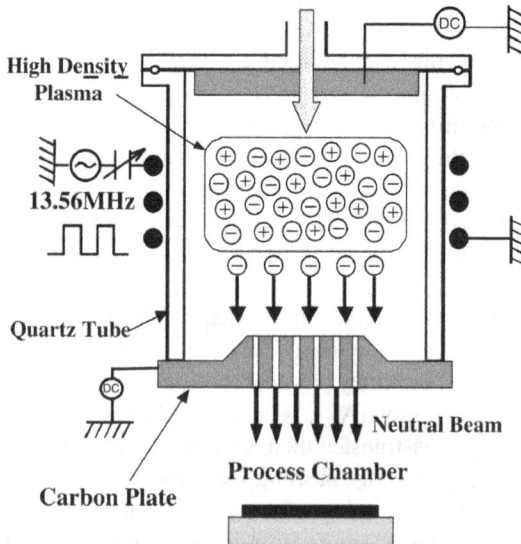

FIGURE 2.5 The newly developed neutral beam etching system [34].

- Coupling: As seen in Figure 2.5, IBE takes place in a triode setup where plasma source is decoupled from substrate, contrary to RIE, for example, where substrate sits inside the plasma glow itself.
- Pressure: In IBE, ions are accelerated in low pressure—around 10^{-4} Torr range [34] —since etching in this case is mainly mechanical and requires preserving as much of the ions' energy to be used in bombarding the substrate's surface. At such pressure, the ions exhibit free molecular flow behavior which means their MFP is large. Sequentially, it means that ions move in nearly straight lines from their source to the substrate.
- Ions energy: When performing IBE to substrates, ions must have energy high enough to break bonds at the surface and sputter atoms off the material which usually happens around 100–500 eV. However, if ion energy becomes even higher, it will start deeply penetrating the surface and turning into a case of implantation or it will cause defects such as lattice damage or point defects [35, 36].
- Selectivity: Since IBE is based on physical etching, it removes all materials similarly without regard to chemical reactivity. Thus, it results in selectivity of nearly 1:1 in the shown equation, which means that selectivity is relatively low in IBE.

$$\text{Selectivity} = \frac{\text{Etch rate of target material}}{\text{Etch rate of underlying material}} \qquad (2.4)$$

- Isotropy: IBE has a high degree of anisotropy which means ions remove the material in a specific direction without causing undercuts since ions are focused in a beam that bombards the surface in a specific direction. This makes IBE suitable for applications where a uniform profile is needed.
- Ion species: In IBE, ions should not be of reactive type materials since we need to only physically etch the material. Commonly, Ar^+ is used as an industry standard since it has a very good balance between inertness and control. Moreover, with a molecular weight of 40, Ar^+ is heavy enough to transfer the kinetic energy to the substrate's surface.
- Beam angle (Figure 2.6): Beam angle plays a very critical role in profile control and anisotropy. Angle also controls a very important parameter which is yield. Yield can be defined as atoms removed/incident ion, and maximum yield does not happen at upright zero angle but happens at angles around 60–80 degrees [35]. In Figure 2.6, (c) $\beta = 45°$, (e) $\beta = 0°$, and (g) $\beta = 0°$ followed by IBE at $\beta = 80°$ [36]. It can be seen that at 80°, the etched height is higher than at the other angles. At higher angles, problems like shadowing or non-uniformity can happen.
- Redeposition: Redeposition of the sputtered material is a problem in IBE. Because etching is a physical process, atoms that are removed from the surface are pushed out in various directions and can redeposit on nearby surfaces like sidewalls. This is particularly strong with shallow beam angles or etching dense patterns. It can lead to profile distortion, passivation, or etch stop for shallow trenches, and sample rotation is often applied in IBE processing to minimize this effect [1].

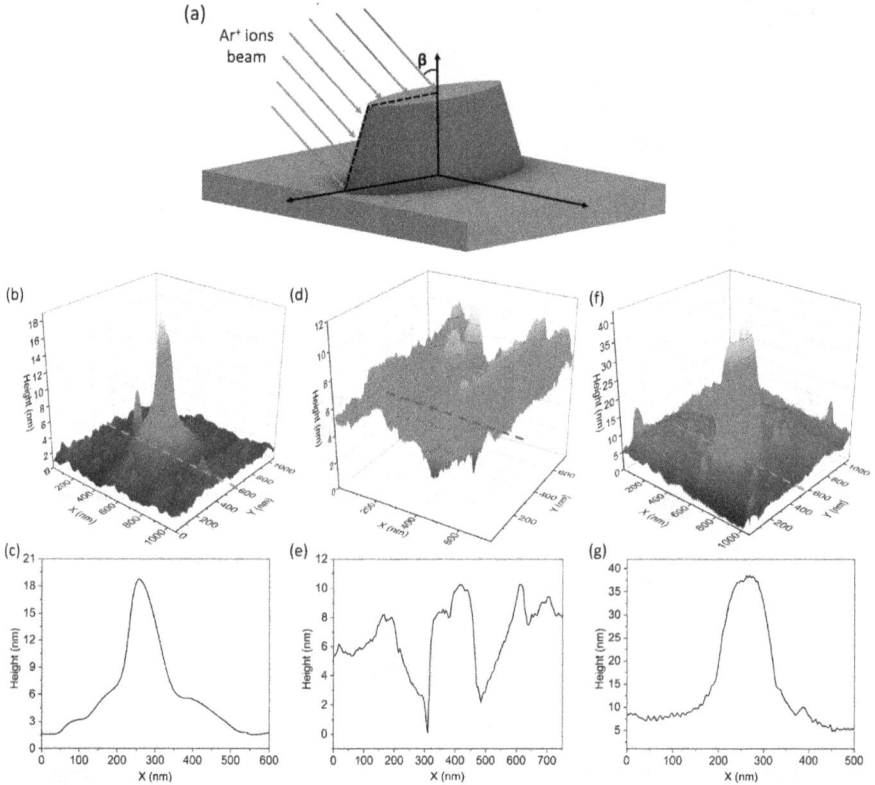

FIGURE 2.6 (a) Convention used to define the IBE angle β. (b) AFM images of the MTJs after IBE at (b) $\beta = 45°$, (d) $\beta = 0°$, and (f) $\beta = 0°$ followed by IBE at $\beta = 80°$. The cross-sectional profiles of AFM images (indicated by the dashed green lines) for (c) $\beta = 45°$, (e) $\beta = 0°$, and (g) $\beta = 0°$ followed by IBE at $\beta = 80°$ [36].

In semiconductor and electronics device failure analysis and characterization, IBE is used in cross-sectional sample preparation for microscopic examination. IBE offers nanometer-level precision for material removal, making it possible to expose regions of interest that are desired with very little damage. Such precision aids in proper defect analysis and quality control [36].

Magnetic devices (hard disk drives): During the fabrication of magnetic memory, IBE enables the accurate patterning of magnetic films or multilayers without chemical damage [37].

In photonics, IBE is used for producing high-accuracy components like gratings or waveguides. For instance, research on barium titanate (BTO) for integrated photonics has shown that IBE can produce smooth sidewalls and accurate pattern transfer, which are crucial in minimizing optical losses [38].

Laser assisted etching (LAE) is a maskless, direct-write etching process based on utilizing a concentrated laser beam to modify material surfaces locally by inducing chemical reactions. LAE relies upon thermal effects where the laser heats the substrate and accelerates reaction rates with an etchant or photochemical effects, where photons of the laser, typically ultraviolet, shatter chemical bonds or excite surface

TABLE 2.4

Comparison between Thermal and Photochemical Etching [39, 40]

Feature	Thermal Etching	Photochemical Etching
Energy source	Heat	UV photon energy
Laser type	IR or visible (e.g., Nd:YAG)	UV (e.g., excimer lasers)
Dominant mechanism	Temperature-enhanced reaction rate	Photoactivation / bond-breaking
Etchant type	Liquid or gas (e.g., HF, HNO_3)	Reactive gas (e.g., Cl_2, SF_6)
Selectivity	Temperature-dependent	Wavelength—and material—dependent
Thermal damage risk	Possible (esp. glass, polymers)	Minimal to none

species with limited heating. This enables sub-micron etching resolution, selective material removal, and low substrate damage [39].

In thermal reaction-driven etching, a substrate is locally heated using a laser beam, accelerating the rate of chemical reaction between the substrate's surface and the surrounding etchant fluid. The concentrated laser has a very localized effect, limiting etching to only the laser-exposed area. The steps of thermal reaction-driven etching start with the substrate exposed to a chemical etchant gas or liquid. Then, a focused laser beam is scanned across the surface. And finally, material is removed only in targeted zones.

In photochemical reaction-driven etching, the laser provides photon energy (particularly in the UV) that: breaks chemical bonds, excites etchant molecules or surface atoms, and generates reactive species (radicals), which leads to chemical reactions that selectively remove material without generating significant heating [40]. In this process, the material is first exposed to a reactive etchant gas. A UV laser is then focused onto the substrate. As a result, material is selectively removed only in the regions illuminated by the laser, enabling precise patterning. Table 2.4 shows the comparison of thermal and photochemical etching [39, 40].

Based on the previous discussion, a comparison can be made between both laser assisted techniques:

APPLICATIONS

- Microfluidic channels: Used for maskless etching of microchannels, cavities, or trenches [41].
- Surface cleaning: Used in removing oxides or photoresist residues.
- Patterning polymers: UV-laser-assisted etching removes or shapes photoresists (e.g., PMMA, SU-8) without masks as needed in LIGA process [42].
- Micro-lens arrays: Close-packed micro-lens arrays (MLA) are widely used in semiconductors and display devices [43].

2.4 FACTORS INFLUENCING ETCHING EFFICIENCY AND SURFACE QUALITY

The efficiency and surface quality of MEMS etching processes are dependent on several physical and chemical factors, since etching is both a chemical and a physical process. As there is advancement in technology, MEMS devices grow more complex, and their structural requirements grow more precise. Hence, scientists and manufacturers need control over etch behavior to achieve reliable and reproducible results. Etch rate control to surface defect mitigation, every step of the etching process must be fine-tuned to optimize the outcome and reduce failures in production and research.

One of the primary and most important considerations while etching is the etch rate, which defines how quickly the material is removed from the substrate. An optimal etch rate ensures that the material is not over-etched, avoiding damage to critical features. In etching, it is compulsory to note the etch rates of all the materials involved to prevent unnecessary etching. It is governed by several factors such as the etchant species, ion energy (in dry processes), and chemical reactivity of the material being etched. In dry etching, adjusting RF power, pressure, and bias voltage can help regulate the rate precisely [44].

Anisotropy, already discussed in detail in Section 2.1, plays a key role in determining the shape and quality of the etched structure. Here, the focus is on how different levels of anisotropy influence etching outcomes, especially in terms of efficiency and smoothness. For example, Bosch DRIE, which is a highly anisotropic technique, is often used to achieve straight vertical sidewalls in deep trenches. On the other hand, if the etching is isotropic, material is removed in all directions, which can lead to undercutting and less defined features. Techniques such as DRIE make use of alternating etch and passivation cycles, e.g., the Bosch process, to create highly anisotropic trenches with etch selectivity as high as 200:1 between silicon and photoresist masks [37]. By alternating between SF_6 etching and C_4F_8 passivation, vertical profiles can be achieved while controlling lateral etching. For many applications, cryogenic etching with SF_6 and O_2 can result in even smoother sidewalls, although it requires precise thermal management at temperatures as low as $-100°C$ [45].

The aspect ratio is the depth-to-width ratio of an etched structure and has a direct effect on etch uniformity and resolution. As the aspect ratio increases, challenges such as aspect ratio-dependent etching (ARDE) or "etch lag" emerge. This phenomenon arises when narrow trenches etch more slowly than wider ones due to restricted reactive species transport or byproduct removal. It can be minimized by adjusting pressure, increasing ion directionality, or modifying gas flow ratios [46].

Surface roughness majorly defines the mechanical and optical performance of MEMS devices. In plasma-based etching, roughness often results from ion-induced damage, micro-masking, or polymer accumulation. For instance, in DRIE, the C_4F_8 passivation layer can cause scalloping on the sidewalls if not carefully controlled. Scalloping refers to the small, periodic sidewall undulations or "ripples" that appear on etched features during the Bosch DRIE process. This occurs because the Bosch method alternates between etching with SF_6 and passivation with C_4F_8, in discrete steps. Each cycle etches a small vertical segment, and the transition between cycles creates a step-like texture on the sidewalls. Scalloping affects surface smoothness and can reduce the functional reliability of moving parts in MEMS actuators or resonators [47]. Techniques such as cycle smoothing, pulsed power modulation, and post-etch treatments like isotropic plasma cleaning are employed to mitigate this issue.

Another factor impacting surface quality is redeposition, which implies the reattachment of etched materials back onto the substrate. This occurs in physical etching or during plasma sputtering processes when ejected atoms do not leave the chamber efficiently and stay near. Redeposition can block trenches, damage resist patterns, or introduce local roughness. Etch chamber design, including directional gas flow, high vacuum efficiency, and periodic in-situ cleaning protocols, are employed for mitigating redeposition effects [48].

Etch-induced damage refers to the defects introduced by high-energy ion bombardment or electrical charging during plasma exposure. When high-energy ions bombard the MEMS device surface, some defects may appear as micro-cracks or stress-induced delamination, particularly in layered MEMS structures. Adjusting plasma density, reducing RF bias, and using low-damage chemistries like Cl_2/HBr instead of F-based plasmas for certain materials can reduce this damage [49].

Temperature strongly affects the etching outcomes, particularly in thermally-assisted and wet chemical etching processes. In wet etching, increased temperature means faster reaction kinetics, improving diffusion rates and leading to faster etch rates. However, overly high temperatures (70–90°C) can lead to non-uniformity, bubble formation, or uncontrolled undercutting in anisotropic wet etchants like Potassium Hydroxide (KOH) or Tetramethylammonium Hydroxide (TMAH) when etching silicon. In contrast, low-temperature plasma etching offers better control over lateral dimensions but may require longer etch durations [50, 51]. Bubble formation in wet etching refers to the creation of gas bubbles on the surface of the wafer during a chemical reaction between the etchant and the substrate. These gas bubbles may stick to the wafer surface, blocking the etchant from reaching certain regions, leading to non-uniform etching, defects, or roughness.

Pressure plays a critical role in determining the behavior of etching species. In dry etching, lower pressures facilitate long MFPs, allowing ions to retain energy and directional momentum. This results in anisotropic etching that can be ideal for narrow trenches or vias [52]. At higher pressures, the increased number of collisions increases isotropy, beneficial for bulk removal but results in poor vertical precision. Pressure also influences the formation of passivation layers, as seen in the Bosch process, where slight changes in chamber pressure can affect polymer coverage and trench profile quality.

The composition of the etching gas directly affects etch selectivity, profile shape, and surface damage. For example, fluorine-based gases like SF_6, CF_4, and CHF_3 are widely used in silicon etching due to their high reactivity with silicon atoms. Adding oxygen in the process helps regulate polymer buildup by combustion, while argon is introduced for its sputtering (bombarding) effect in physical etching. For materials like SiO_2 or Si_3N_4, fluorocarbon gases such as C_4F_8 provide good selectivity, whereas Cl_2 and BCl_3 are preferred for metals like Al and Ti [53]. Selectivity is the ability to etch the target material while leaving the mask or underlying layers intact. Scientists may change gas mixtures, flow rates, and power settings to enhance selectivity, depending on the materials in the wafers. This is important when delicate photoresists or thin hard masks are used, as their erosion would compromise the overall pattern production. Selectivity may not promise the prevention of etching these photoresists or masks, but it will slow down the etching process of these substances, targeting only the materials that need to be etched. Longer exposure time of etching may result in unwanted etching of these materials; hence, the time of etching

TABLE 2.5

Comparison of Etching Techniques Based on Operating Parameters and Surface Characteristics [54, 55]

Etching Method	Typical Temperature	Pressure Range	Etch Rate (Si)	Anisotropy	Surface Roughness	Selectivity (Si:Mask)
Bosch DRIE	20–40°C	20–60 mTorr	5–20 μm/min	Very high	Moderate (scalloping)	75:1 (PR), 150:1 (SiO_2)
Cryogenic DRIE	–100°C (173 K)	10–30 mTorr	4–5 μm/min	High (smooth sidewalls)	Low (optically smooth)	100:1 (SiO_2 mask)
Wet Etching (KOH)	70–90°C	Atmospheric	1–2 μm/min	Low (isotropic)	Low to moderate	30–50:1 (mask dependent)

also plays a vital role in the etching process. In summary, we show a comparison of etching techniques based on operating parameters and surface characteristics in Table 2.5 [54, 55].

REFERENCES

1. Madou, M. J. (2011). *Fundamentals of microfabrication and nanotechnology* (3rd ed., vols. 1–3). CRC Press.
2. Franssila, S. (2010). Etching. In *Introduction to microfabrication* (pp. 127–142). John Wiley & Sons Ltd.
3. Franssila, S. (2010). Anisotropic wet etching. In *Introduction to microfabrication* (pp. 237–254). John Wiley & Sons Ltd.
4. Nayak, A. P., Islam, M. S., & Logeeswaran, V. J. (2012). Wet etching. In B. Bhushan (Ed.), *Encyclopedia of nanotechnology*. Springer.
5. MicroChemicals GmbH. (2025). Wet-chemical etching of silicon and SiO_2. Retrieved from https://www.microchemicals.com
6. MicroChemicals GmbH. (2025). Wet chemical etching – Basics. Retrieved from https://www.microchemicals.com
7. Franssila, S. (2010). Deep reactive ion etching. In *Introduction to microfabrication* (pp. 255–270). John Wiley & Sons Ltd.
8. Nayak, A. P., Islam, M. S., & Logeeswaran, V. J. (2012). Dry etching. In B. Bhushan (Ed.), *Encyclopedia of nanotechnology*. Springer.
9. Williams, K. R., Gupta, K., & Wasilik, M. (2003). Etch rates for micromachining processing—Part II. *Journal of Microelectromechanical Systems*, 12, 761–778. https://doi.org/10.1109/jmems.2003.820936
10. Fogler, H., Goldsmith, B., & Singh, N. (2025). Rate laws. In *Elements of chemical reaction engineering* (7th ed., Chapter 3). Pearson Education, Inc.
11. Katopodes, N. D. (2019). Diffusive mass transfer. In *Free-surface flow* (pp. 184–270). https://doi.org/10.1016/b978-0-12-815489-2.00003-4
12. Swenson, H., & Stadie, N. P. (2019). Langmuir's theory of adsorption: A centennial review. *Langmuir*, 35, 5409–5426. https://doi.org/10.1021/acs.langmuir.9b00154
13. Ossei-Wusu, E., Carstensen, J., Quiroga-González, E., et al. (2013). The role of polyethylene glycol in pore diameter modulation in depth in p-type silicon. *ECS Journal of Solid State Science and Technology*. https://doi.org/10.1149/2.020305jss

14. Kistemaker, J. C., Lubbe, A. S., Bloemsma, E. A., & Feringa, B. L. (2016). On the role of viscosity in the Eyring equation. *ChemPhysChem*, 17, 1819–1822. https://doi.org/10.1002/cphc.201501177

15. City University of Hong Kong. (2025). Silicon etching. Retrieved from https://www.cityu.edu.hk/phy/appkchu/AP6120/6.PDF

16. MicroChemicals GmbH. (2025). Silicon etching. Retrieved from https://www.microchemicals.com/dokumente/application_notes/silicon_etching.pdf

17. IQS Directory. (2025). Acid etching: Types, applications and benefits. Retrieved from https://www.iqsdirectory.com/articles/metal-etching/acid-etching.html

18. IDC Technologies. (2025). Acid-based etching of wafers. Retrieved from https://www.idc-online.com/technical_references/pdfs/chemical_engineering/Acid_Based_Etching_of_Wafers.pdf

19. Kong, L., Dasgupta, B., Ren, Y., Mohseni, P. K., Hong, M., Li, X., Chim, W. K., & Chiam, S. Y. (2016). Evidences for redox reaction driven charge transfer and mass transport in metal-assisted chemical etching of silicon. *Scientific Reports*, 6, 36582. https://doi.org/10.1038/srep36582

20. Romano, L., Lawley, C., Halvorsen, M. M., Jefimovs, K., Guzenko, V., & Stampanoni, M. (2023). Metal-assisted chemical etching: Towards CMOS compatible catalyst for high aspect ratio nanostructures. In International Conference on Electron, Ion, and Photon Beam Technology and Nanofabrication (EIPBN). Retrieved from https://eipbn.org/abstracts/2023/papers/8C-2.pdf

21. Buzzi, F. L., Ting, Y. H., & Wendt, A. E. (2009). Energy distribution of bombarding ions in plasma etching of dielectrics. *Plasma Sources Science and Technology*, 18(2), 025009. https://doi.org/10.1088/0963-0252/18/2/025009

22. Hori, M. (2022). Radical-controlled plasma processes. *Reviews of Modern Plasma Physics*, 6, 36.

23. Kim, D. S., Jang, Y. J., Kim, Y. E., Gil, H. S., Kim, H. J., Ji, Y. J., Kim, H. Y., Kim, I. H., Chae, M. K., Park, J. C., & Yeom, G. Y. (2022). Radical flux control in reactive ion beam etching (RIBE) by dual exhaust system. *Applied Surface Science*, 571, 151311.

24. Medical News Today. (2023). Fluoride: Risks, uses, and side effects. Retrieved from https://www.medicalnewstoday.com/articles/154164

25. Kim, J., Lee, S., Park, H., et al. (2024). Fluoride removal from wastewater and potential for resource recovery: Comparative studies between different treatment technologies. *Environmental Engineering Research*, 29(2), 179.

26. Coburn, J. W., & Winters, H. F. (2001). Silicon etching yields in F_2, Cl_2, Br_2, and HBr high density plasmas. *Journal of Vacuum Science & Technology A*, 19(5), 2197–2201.

27. Mayer, A. (2002). Plasma etching in microelectronics. NASA Technical Reports Server (NTRS). Retrieved from https://ntrs.nasa.gov/api/citations/20020059586/downloads/20020059586.pdf

28. Zhang, J., Zhang, F., Ma, M., & Liu, Z. (2024). Fabrication of ordered macropore arrays in n-type silicon wafer by anodic etching using double-tank electrochemical cell. *Micromachines*, 15(5), 569.

29. Surdo, S., & Barillaro, G. (2024). Voltage- and metal-assisted chemical etching of micro and nano structures in silicon: A comprehensive review. *Small*, 20(35), e2400499.

30. Wang, X. (2017). Ion beam etching. In M. Aliofkhazraei (Ed.), *Ion beam applications* (pp. 103–123). IntechOpen.

31. Suryana, R., Aini, N. A., & Diantoro, M. (2024). Effect of HF and ethanol volume ratio on porous silicon formed by photoelectrochemical method on n-type Si(100) substrates. *Evergreen*, 11(3), 2127–2134.

32. Kuntyi, O., Zozulya, G., & Shepida, M. (2022). Porous silicon formation by electrochemical etching. *Advances in Materials Science and Engineering*, 2022, 1482877.

33. Bhattacharyya, B. (2015). Design and developments of microtools. In *Electrochemical micromachining for nanofabrication, MEMS and nanotechnology* (pp. 101–122). William Andrew Publishing. https://www.sciencedirect.com/science/article/abs/pii/B9780323327374000062

34. Samukawa, S. (2006). Ultimate top-down etching processes for future nanoscale devices: Advanced neutral-beam etching. *J. Appl. Phys*, 45, 2395.

35. Ahrens, P., Zander, M., Hirsch, D., Hasse, U., Wulff, H., Frost, F., & Scholz, F. (2019). Influence of argon ion beam etching and thermal treatment on polycrystalline and single crystal gold electrodes Au(100) and Au(111). *Journal of Electroanalytical Chemistry*, 832, 233–240.

36. Ren, H., Wu, S.-Y., Sun, J. Z., & Fullerton, E. E. (2021). Ion beam etching dependence of spin–orbit torque memory devices with switching current densities reduced by Hf interlayers. *APL Materials*, 9(9), 091101.

37. Yang, L., Chen, E. Y.-S., Cairney, J. M., Qu, J., Garbrecht, M., McCarroll, I. E., Mosiman, D. S., & Saha, B. (2025). Improved atom probe specimen preparation by focused ion beam with the aid of multi-dimensional specimen control. *Microstructures*, 5, 2025007.

38. Zhao, M., Liu, Y., Ma, R., Qiu, Y., Zhao, X., Zheng, S., Zhong, Q., Dong, Y., & Hu, T. (2025). Ion beam etching of barium titanate for integrated photonics. *Journal of Vacuum Science & Technology A*, 43(3), 032601.

39. Cui, Z. (2024). Nanoscale pattern transfer by etching. In *Nanofabrication* (pp. 257–297). Springer International Publishing.

40. Liu, H., Lin, W., & Hong, M. (2021). Hybrid laser precision engineering of transparent hard materials: Challenges, solutions and applications. *Light: Science & Applications*, 10, 162.

41. Liu, X.-Q., Bai, B.-F., Chen, Q.-D., & Sun, H.-B. (2019). Etching-assisted femtosecond laser modification of hard materials. *Opto-Electronic Advances*, 2, 190021.

42. Chalker, P. R. (2016). Photochemical atomic layer deposition and etching. *Surface and Coatings Technology*, 286, 2–7.

43. Kim, S., Kim, J., Joung, Y.-H., Ahn, S., Choi, J., & Koo, C. (2019). Optimization of selective laser-induced etching (SLE) for fabrication of 3D glass microfluidic device with multi-layer micro channels. *Micro and Nano Systems Letters*, 7, 94.

44. Srikantaprasad, G., Mathew, N., & Nagar, S. (2024). Laser micromachining on PMMA: An efficient fabrication of microchannels for sustainable microfluidic devices. *Journal of the Brazilian Society of Mechanical Sciences and Engineering*, 46, 4904.

45. Baek, G. H., Hwang, E. S., & Cheong, B.-H. (2022). Quasi-periodic micro-lens array via laser-assisted wet etching. *AIP Advances*, 12(10), 105219.

46. Huff, M. (2021). Recent advances in reactive ion etching and applications of high-aspect-ratio microfabrication. *Micromachines*, 12(8), 991.

47. Laermer, F., Schilp, A., Funk, K., & Lang, V. (1999). Bosch deep silicon etching: Improving uniformity and etch rate for advanced MEMS applications. In *IEEE International MEMS Conference*.

48. Xu, J., et al. (2024). Deep-reactive ion etching of silicon nanowire arrays at cryogenic temperatures. *Applied Physics Reviews*, 11(2), 021101.

49. Ayon, A. A., Braff, R. A., Lin, C., Sawin, H. H., & Schmidt, M. P. (1999). Characterization of a time multiplexed inductively coupled plasma etcher. *Journal of the Electrochemical Society*, 146(1), 339–349.

50. Wu, B., Kumar, A., & Pamarthy, S. (2010). High aspect ratio silicon etch: A review. *Journal of Applied Physics*, 108(5), 051101.

51. Chen, K. S., Ayon, A. A., Zhang, X., & Spearing, S. M. (2002). Effect of process parameters on the surface morphology and mechanical performance of silicon structures after deep reactive ion etching (DRIE). *Journal of Microelectromechanical Systems*, 11(3), 264–275.

52. Rangelow, I. W. (2003). Critical tasks in high aspect ratio silicon dry etching for microelectromechanical systems. *Journal of Vacuum Science & Technology A*, 21(4), 1550–1562.
53. Xu, T., Tao, Z., Li, H., Tan, X., & Li, H. (2017). Effects of deep reactive ion etching parameters on etching rate and surface morphology in extremely deep silicon etch process with high aspect ratio. *Advances in Mechanical Engineering*, 11(4), 1687814017738152.
54. Pal, P., & Sato, K. (2015). A comprehensive review on convex and concave corners in silicon bulk micromachining based on anisotropic wet chemical etching. *Micro and Nano Systems Letters*, 3, 6.
55. Rangelow, I. W. (2001). Dry etching-based silicon micro-machining for MEMS. *Vacuum*, 62(2–3), 279–291.

3 Fluorine-Free Dry Etching for MEMS

3.1 CHLORINE-BASED PLASMA ETCHING FOR MEMS

Chlorine-based plasma etching is a fluorine-free alternative suitable for silicon and oxide materials, but requires careful process control to avoid metal corrosion. This technique chemically interacts with the substrate by using reactive chlorine species produced in plasma to produce volatile by-products that are effectively evacuated, producing distinct etch profiles [1]. Commonly used gases in this procedure include chlorine (Cl_2), boron trichloride (BCl_3), and hydrogen chloride (HCl).

Since chlorine-based plasma etching can achieve high aspect ratios while maintaining material integrity, it is essential to produce a variety of micro-electro-mechanical systems (MEMS) devices, including sensors, optical components, and microfluidic structures. According to a study by Tian et al., despite chlorine-based plasma etching methods having slower etch rates of around 112 nm/min compared to around 195 nm/min for fluorine-based techniques, chlorine-based etching has superior selectivity and less lateral etching, which improves dimensional and profile control, especially in high-aspect-ratio MEMS systems [2].

However, to guarantee the effectiveness and dependability of the etching process, issues like metal corrosion and by-product control need careful process optimization and the selection of suitable chamber materials. Chlorine (Cl_2) gas has a high chemical reactivity with a variety of materials, including AlN, Pt, and SiC, making it a major etchant in plasma-based dry etching for MEMS. Figure 3.1a shows that a flow rate of 40 sccm of Cl_2 yields around 110 nm/min etching rate, making it useful in high-aspect-ratio MEMS structures.

When Cl_2 dissociates in plasma, it produces chlorine ions (Cl^+) and radicals ($Cl\bullet$), which react with the substrate surface to produce volatile metal chlorides such as

(a)   (b)

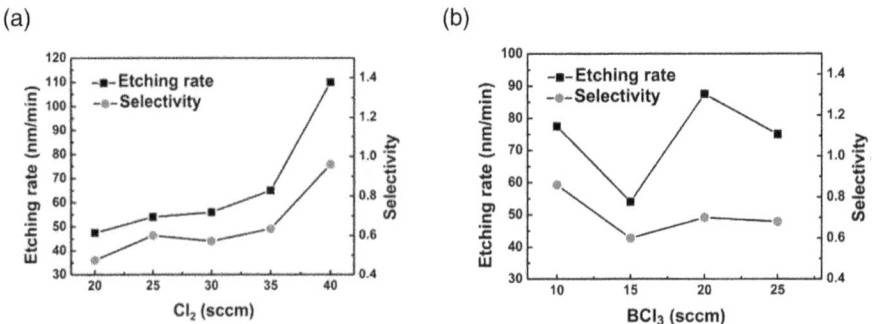

FIGURE 3.1 (a) Aluminum nitride (AlN) etching rate and selectivity versus the flow rate of Cl_2. (b) AlN etching rate and selectivity versus the flow rate of BCl_3 [3].

                                          DOI: 10.1201/9781003674795-3

$PtCl_2$ and $AlCl_3$. Then, these volatile by-products are effectively removed, enabling high etch speeds and clear profiles [2, 3]. Cl_2 has good selectivity and etch control, making it particularly useful for etching platinum and aluminum nitride (AlN). It can also be used for directional ion bombardment by applying a bias voltage during plasma etching, which helps remove material more vertically than sideways, an effect known as anisotropy resulting in cleaner and more precise feature profiles [3].

In MEMS plasma etching, BCl_3 is commonly used together with Cl_2 to maximize surface quality and enhance etch stability. As shown in Figure 3.1b, a flow rate of 20 sccm of BCl_3 yields around 85 nm/min etching rate. It helps to eliminate native oxides and any remaining moisture from the substrate surface and etch chamber, resulting in more uniform and smoother etched surfaces [3]. Furthermore, BCl_3 can improve material homogeneity during deep etching by influencing etch behavior at grain boundaries [4]. Although BCl_3 has a less direct etching contribution than Cl_2, it is nevertheless essential for producing high-quality, repeatable etch results in intricate MEMS systems [4].

In plasma etching, HCl gas is typically utilized for copper (Cu) patterning in microelectronic applications. Cu and reactive chlorine species combine in HCl-based plasma to generate the non-volatile $CuCl_x$ compound, which sticks to the surface after it is etched. To finish the etching process, this chemical is then mixed in a diluted HCl solution, allowing photoresist masks to precisely define the Cu line [5]. The process uses both ion bombardment and plasma-phase chemistry, with higher bias voltages and substrate temperatures improving Cu consumption and etch profile control [4]. This technique has been shown to be effective in IC production, but its great accuracy and material compatibility point to possible MEMS applications. Furthermore, during poly-Si etching, HCl can occur as a by-product in Cl_2/HBr plasmas and be tracked in real-time using diode laser absorption spectroscopy to regulate plasma temperatures and etch rates [6].

MEMS sensors frequently use chlorine-based plasma etching to precisely print materials like silicon and platinum. Choi et al. demonstrated a process for patterning platinum electrodes on MEMS substrates using a Cl_2-based reactive ion etching (RIE) technique, producing smooth, vertical profiles that are crucial for sensor performance [4]. According to a study by Paul et al., BCl_3 and Cl_2 gas mixtures in conjunction with regulated ion bombardment allow for anisotropic etching and the creation of high aspect-ratio structures in silicon micromachining, which is essential for the diaphragms and beams in pressure sensors and accelerometers [7]. Thus, selective etching and decreased undercut are made possible by chlorine-based plasmas, which are essential when designing dependable MEMS sensor geometries.

It is crucial to achieve smooth and precisely angled features in optical MEMS devices, including tunable lenses or micro-mirrors. Paul et al. noted that deep silicon etching, which is essential for optical path alignment, can be accomplished by tuning alternating BCl_3 and Cl_2 flows to regulate sidewall angles [7]. Additionally, Petit-Etienne et al. examined how surface reactions and ion energy impact the smoothness and etch profile of thin SiO_2 layers, which are frequently employed as optical insulators or reflective interfaces, in response to Cl_2 plasma [8].

Microfluidic channel creation in silicon and other materials is also supported by chlorine-based plasma etching. Cl_2-based ICP etching for silicon carbide (SiC) was examined by Racka-Szmidt et al., who emphasized its use in harsh-environment microfluidic MEMS applications [9]. Although molding and embossing are frequently used to create polymer microfluidics, Becker and Gärtner highlighted that hybrid systems that include silicon or metal features into polymer chips also use dry etching, employing chlorine plasma [10].

Chlorine-based plasma etching results in unstable and polymeric by-products that must be properly handled to avoid wafer contamination. Particle contamination and reduced yield can result from the accumulation and redeposition of chemical by-products such as $AlCl_3$, which is produced during the etching of Al using Cl_2/BCl_3 [11]. Additionally, the composition of the by-product is greatly influenced by the passivation gas selection; ethylene diluted in helium (C_2H_4/He) produces thinner, more readily removable polymer coatings, whereas nitrogen (N_2) produces thick, hard polymers that trap chlorine residues [12]. To control particle contamination and surface residues, these findings highlight the necessity of customized gas chemistries and strict chamber maintenance.

One of the biggest problems following aluminum (Al) etching using chlorine-based plasmas is corrosion. In damp conditions, residual $AlCl_3$ on sidewalls can hydrolyze and produce HCl, which feeds the corrosion cycle [12]. According to Yoo et al., the development of theta-phase (Al_2Cu) precipitates, which act as local cathodes in galvanic corrosion cells and result in severe pitting after wet treatment, make Al-Cu alloys particularly prone to corrosion [13]. However, developer-based cleaning solutions may also accelerate pitting, particularly at the borders between Al and barrier layers even while they successfully remove residual chlorine [13]. Additionally, Lee et al. discovered that Cl residues are somewhat higher with N_2 passivation than with C_2H_4/He, establishing a direct correlation between corrosion hazards and gas chemistry [12]. To prevent electrochemical degradation without compromising residue removal, it is crucial to use the best post-etch cleaning procedures.

While chlorine gas is effective in plasma etching processes, it does possess significant environmental and health challenges. Because of its potent pulmonary irritating qualities, chlorine gas presents serious health risks. It forms hydrochloric and hypochlorous acids in the respiratory tract when inhaled, which damage epithelial cells and can cause pulmonary edema and acute respiratory distress syndrome (ARDS). Concerns about safety are increased by the fact that chlorine is a chemical weapon due to its availability and toxicity.

Strict safety procedures are necessary in microfabrication etching operations to reduce the dangers of exposure to chlorine gas. To avoid skin and eye contact with chlorine, personnel should use the proper personal protective equipment (PPE), such as safety goggles, protective clothes, and gloves that can withstand chemicals. Self-contained breathing apparatus (SCBA) use is advised as respiratory protection is vital [14].

3.2 BROMINE-BASED ETCHING FOR MEMS SENSORS AND RF DEVICES

Bromine-based etching process causes less plasma damage than the fluorine-based etching. Regarding fluorine-based etching on Si, it has extremely high corrosion resistance but low orientation [15]. This is due to the small size of the fluorine atom, allowing it to penetrate the surface of the material and causing a spontaneous chemical reaction, forming saturated volatile fluorine silicon compounds [15, 16]. Although this reaction accelerates the corrosion rate, it also leads to lateral corrosion, causing erosion of the side walls and damage to the underlying structures. Meanwhile, bromine only reacts with a limited number of bonds, and the etching process requires ion bombardment on the surface to desorb unsaturated, low-volatility compounds. This prevents lateral etching, making it ideal for trench or gate etching with vertical sidewalls and flat bottoms [16–19]. Besides, the bromine-based etching rate is pretty slow, much slower compared to the fluorine one, but offers higher selectivity.

Similarly, for SiO_2, the research of Efremov [20] and Pearton [21] also shows the same results: Br_2, even with the assistant of ion, still records a slow rate but provides high anisotropy, lower plasma damage, and smoother etched surface in return. And the consistent results were also observed when using Br_2 etching on III-V materials, specifically GaAs [21, 22]. A notable point about this case is that Bromine-based etching can create a smooth surface in the low-pressure condition.

Overall, based on the research results, it can be summarized that the characteristics of Br_2 etching method when use with Si, SiO_2, and GaAs are lower etching rates, smoother etched surfaces, higher anisotropy and selectivity, low plasma damage.

With those characteristics of the method using Br_2, it is a suitable method for etching RF Mems switches, photonic Mems, and GaAs-based devices. Firstly, in the applications for RF MEMS switches, with high anisotropy and precise selectivity, the Br-based method allows accurate pattern definition etching without deformation, avoiding issues such as bridge collapse or stiction. This process also helps reduce residual stress, which can affect the underlying layers and lead to warping or other deformations in the structure. Otherwise, these shape changes can influence the air gap capacitance and consequently impact the device's static characteristics such as pull-in voltage, dynamic behavior like resonant frequency, switching time, and RF performance with insertion loss, isolation as instances [23]. Moreover, if the using method is based on fluorine, it would be easy leading to rough surfaces and residue buildup, which are the reasons for the increase in capacitance, electric field concentration, and leakage current factors that promote stiction [24, 25]. In the opposite, the smooth surfaces created by Br-based etching can effectively eliminate these risks.

Lastly, the abovementioned advantages are also applicable to the etching processes used in photonic MEMS and GaAs devices. Additionally, the smooth surface is extremely necessary to create micro mirror, which is a key component in MOEMS and photonic MEMS applications [18]. Although having notable advantages that make it suitable for specific applications, bromine-based etching still has to face several limitations. The most significant one is the slow etching rate, which is repeated continuously on research articles [16–22]. Therefore, the research of Vitale et al. [16] points out that the slow etching rate is due to the low ion flux but not because of the etching yield. In the application for SiO_2, the slow volatility speed of the reaction products is also a challenge [21]. Specifically, the main product of the reaction between HBr plasma and SiO_2 is $SiBr_4$, which has a boiling point of 154°C. This makes it difficult to evaporate, causing it to remain on the surface and hinder the etching process. Facing these limitations, some methods can be considered in order to increase the etching rate. For example, we can increase the ion flux as it is one of the main reasons for the slow speed of this process. Moreover, the temperature of the silicon surface is also a significant factor when the etching rate decreases significantly when the temperature of silicon surface is over 60°C [16] as at this condition, the surface coverage by adsorbed Br and H atoms decreases. This suggests that the suitable temperature should be kept lower than 60°C. Another method also has potential to be considered is applying the combination of different chemicals in etching process, like Cl, in order to accelerate the etching speed. The research conducted by Vitale et al. [16] summarizes that the combination of Cl_2 and HBr improves the etching speed and efficiency, more than the methods only using HBr or Cl_2 separately. At the same time, it helps to decrease the sidewall and microtrenching phenomenon that usually be seen when using pure Cl_2. Additionally, the combination gives a stable condition that eliminate unexpected deformation when create microstructure in silicon.

Besides, the toxicity and high corrosion characteristics of bromine-containing gases used in etching process like Br_2, HBr, and CF_3Br also proposes significant concerns, as they are not only dangerous for human health but also can harm and reduce the lifespan of the equipment [22]. Therefore, it is important to have suitable treatment with those chemicals, as well as a suitable requirement in working process to control the gases, eliminate the harmful effects to human and tools.

3.3   RADICAL-ASSISTED DRY ETCHING FOR HIGH-PRECISION MEMS

Radical-assisted dry etching (RADE) is an alternative dry etching technique that eliminates the dependency on fluorine-containing ions and radicals. As the name suggests, radicals are generated using plasma, but the etching process primarily relies on chemical reactions rather than physical ion bombardment. The method has been increasingly utilized since the 2000s, driven by its highly selective and less destructive etching properties.

Plasma is generated in a separate chamber from the wafer (known as remote plasma), where gases such as O_2, H_2, or Cl_2 are exposed to low-energy plasma conditions that favor the formation of neutral radicals rather than ionized species. The neutral radicals are transported to the main chamber through gas flow systems. Ions and electrons recombine naturally or are filtered out using grids or magnetic fields that block or deflect them. The neutral radicals are highly reactive and chemically interact with the wafer surface, forming volatile compounds that can be easily removed by the vacuum system [26]. O_2, H_2, and Cl_2 are all gases used in RADE, but each is preferably selected based on the specific material and etching application.

Oxygen radicals are mainly used for etching away carbon-based materials. They react with these materials, breaking chemical bonds and forming volatile by-products such as carbon monoxide (CO) and carbon dioxide (CO_2), which can then be removed from the system. This makes oxygen radicals particularly effective for removing photoresist layers, polymers, and other organic contaminants from surfaces. As shown in Figure 3.2a, this process involves the interaction between oxygen radicals and the carbon-based material, leading to the formation of gaseous reaction products that are evacuated during the etching step.

Alternative methods used in the industry to remove photoresist, or organic contaminants include soaking in chemical bath, thermal ashing using high-energy O_2 plasma in direct contact with the wafer, and UV-ozone cleaning, where ultraviolet light breaks down organics in the presence of ozone (O_3) [27].

Some of the advantages of this method are:

- Uses non-toxic, non-corrosive that are also in abundance in nature
- Highly selective—targets organic compounds without affecting underlying layers or materials
- It's a gas phase process, which allows the penetration of deep features, outperforming wet chemical methods
- Avoids sputtering and charging effect

The etch rate for oxygen RADE typically ranges from 300 to 800 nm/min, depending on factors such as radical density, substrate temperature, plasma power, and photoresist composition [27].

Some of the industrial applications that utilize this method include MEMS fabrication, where its non-destructive and highly penetrating nature is ideal for delicate structures; 3D integration, where it is used to clean bonding surfaces by removing organic contaminants; and BioMEMS or microfluidic devices, where it removes organic residues without damaging soft materials and also prepares surfaces for bonding.

Hydrogen radicals are used to remove halogen residues or native oxides and to prepare surfaces for wafer bonding. Unlike oxygen radicals, which break C–C bonds through oxidation, hydrogen radicals function as reducing agents. They react with halogens, oxides, and weak organic compounds, forming volatile by-products such as hydrogen fluoride (HF), HCl, or water vapor. These reaction products are gaseous and are efficiently evacuated from the system by the vacuum pump. As shown in Figure 3.2b, this process highlights how hydrogen radicals interact with surface contaminants and facilitate their removal through chemical reduction.

Wafer-level bonding done in 3D packaging requires very clean surfaces, no oxides, and no halogens to prevent voids. During the reduction the hydrogen radicals

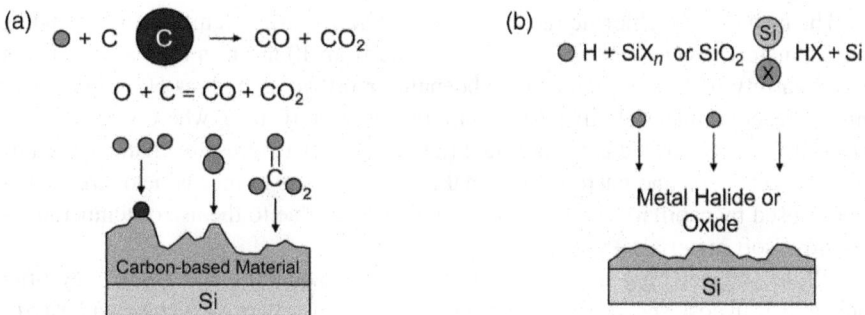

FIGURE 3.2 (a) O radicals react with carbon compounds, forming CO/CO_2 removed by vacuum. (b) H radicals remove halogens and oxides, forming HF, HCl, and H_2O (self-made).

create Si-H bonds removing oxygen in the form of water vapor. The temperature of this process is below the bonding temperature.

The process of creating Si–H bonds is known as passivation [28]. It chemically stabilizes the wafer surface, preventing it from re-oxidizing. While H_2-based RADE offers advantages such as a clean, dry process and the ability to form a passivating Si–H layer, it also has limitations—the etching or cleaning rate is relatively slow, and the use of flammable hydrogen gas raises important safety and handling concerns [28].

Currently, to optimize this technique in industry, research is being conducted on reducing process time, developing surface depassivation methods, and exploring the use of additional radicals in combined processes. Hydrogen radical-assisted etching used for removing native oxides and halogen residues prior to wafer bonding typically exhibits a cleaning rate of 5–30 nm/min [28].

Cl-radical-assisted etching is used in the industry due to the reactivity with Si, GeAs, and metals. For the process gases like Cl_2, BCl_3, or $SiCl_4$ are dissociated into Cl radicals via RF power. These gases are easier to dissociate compared to oxygen or hydrogen—low-energy electrons can generate Cl radicals efficiently. Cl radicals react with the surface of the material (metal or semiconductor) and form metal chlorides or non-metal chlorides.

Some of the byproducts are very volatile, like $SiCl_4$, $TiCl_4$, $GaCl_3$, $AsCl_3$, and PCl_3, making the cleaning easier. By-products like $InCl_3$, $MoCl_6$, and $HfCl_4$ are less volatile, meaning they do not evaporate under operating conditions, which increases the risk of redeposition [29]. This method is used to etch metals such as Al interconnects, titanium barrier layers, and, less commonly, molybdenum metal gates due to the formation of less volatile by-products that are harder to remove.

As mentioned before, it is also used in etching semiconductors like GaAs used in RF, optoelectronics, LEDs [30], and, less commonly, indium phosphide (InP) used in photonics and high-speed logic. For silicon etching, fluorine-based chemistries are generally more practical due to their higher etch rates. However, chlorine radicals can be used in cases where a slower, more controlled process is required. This makes Cl-based etching preferable in applications such as MEMS fabrication, front-end-of-line (FEOL) processing in advanced nodes, and gate stack patterning—where improved sidewall control, reduced undercutting, and higher selectivity over underlying layers are critical [30].

The etch rate for chlorine radical-assisted etching varies significantly depending on the material being etched, typically ranging from 10 nm/min for less reactive or low-volatility materials such as molybdenum or InP, to as high as 500 nm/min for highly reactive materials like Al or gallium arsenide (GaAs), where volatile chlorides are readily formed and evacuated [29]. MEMS structures are characterized by high aspect ratios and narrow gaps, making them prone to mechanical stress, stiction caused by capillary forces, and thermal damage due to the use of temperature-sensitive soft materials.

While other etching methods such as wet etching, RIE, and DRIE may offer advantages in cost, speed, or high aspect ratio capability, they each come with significant shortcomings that can impact delicate MEMS structures. Wet etching cannot effectively penetrate high-aspect-ratio structures and often leads to capillary forces during drying, resulting in stiction. In contrast, RIE and DRIE may cause physical damage from ion bombardment as well as charging effects that can harm delicate MEMS structures.

RADE uses neutral radicals to remove material chemically, avoiding sputtering and physical damage. The process is dry and operates at temperatures below levels that could damage sensitive structures or materials [31]. The absence of sputtering in RADE helps preserve the mechanical integrity of MEMS structures, maintain smooth surfaces, and protect coatings or bio-interfaces when present. This is crucial for retaining key performance parameters such as resonant frequency, sensitivity, and structural reliability [31]. Different from DRIE or RIE that use ion bombardment directionality, radical-assisted dry etching relies on the chemical reaction of the radicals. In RADE factors like radical delivery, surface conditions, or chamber environment directly impact etch rate and uniformity.

Neutral radicals diffuse randomly instead of moving in straight lines, so their distribution in the chamber impacts the etch rate [31]. Additionally, temperature plays a critical role in RADE, since the surface reaction rate is strongly temperature-dependent [31]. Even slight variations in wafer temperature can lead to differences in the rate at which material is removed, ultimately impacting the uniformity of the etch across the wafer.

Strategies used in the industry to mitigate the radical distribution are:

- Reactor showerheads and gas inlets are carefully engineered to deliver a uniform distribution of reactive gases across the wafer surface, ensuring consistent radical generation and transport [32].
- Computational models that simulate how gases (and radicals) move through the etch chamber, and help tune process parameters like pressure and flow rate to improve uniformity [32].

Strategies used in the industry to maintain temperature uniformity are:

- Electrostatic chucks (ESCs) with precise temperature control and uniform backside helium cooling [33].
- Process recipe tuning to reduce heating from radical recombination or exothermic reactions [33].

3.4   PLASMA-FREE METAL-ORGANIC ALE FOR DELICATE MEMS MATERIALS

Atomic layer etching (ALE) is a plasma-free etching method with extremely high precision at atomic or near-atomic level, based on the principle of removing material one atomic layer by layer [34, 35]. By using a series of individual self-limiting surface reactions, the etching depth can be precisely controlled strictly by the number of ALE cycles, which would become critically vital when going down to under 10 nm size [34]. Moreover, an additional advantage of ALE is its potential to provide higher material selectivity, compared to conventional continuous plasma etching methods. It can preferentially etch a specific material while minimizing or eliminating the

TABLE 3.1

A Comparison of Conventional Etching and Atomic Layers Etching Techniques [34, 35]

Aspect	Conventional Etching	Atomic Layer Etching (ALE)
Etch mechanism	Bulk material removal via chemical reactions, physical sputtering, or a combination of both	Cyclic process that alternates between surface modification and activation steps
Etch rate	Ranges from 100 to 1000 nm/min. Depends on factors like gas chemistry, RF power, chamber pressure, substrate temperature, and plasma density	Ranges of 0.1–0.5 nanometers per cycle. Depends on substrate temperature, dosing and purge timing, and chamber conditions such as pressure and flow rate
Precision	Nanometer-scale resolution achieved	Atomic-level precision
Surface roughness	There is surface roughness due to high-energy ions used	No physical damage
Applications	Ultra-thin film control, sensitive surfaces	General patterning

erosion of other materials. At the same time, it significantly reduces surface damage, resulting in smoother etched surfaces [34].

Compared to conventional etching methods, ALE has many advantages which are illustrated by Kanarik [36] by comparing ALE to RIE. To be more specific, conventional etching methods were defined as continuous and simultaneous processes, where many surface reactions occur at the same time, and RIE is currently the most common directional etching technique. It is a fast-etching method that involves hundreds of simultaneous plasma/surface interactions, using complex gas mixtures and high-energy ions (~1000 eV) to achieve selectivity. However, the etch rate tends to be limited by the reactant flux. The uniformity can also be affected by the change of gas flow in the reaction chamber. Moreover, simultaneous deposition and the formation of a selvage layer (5–20 nm) reduce process control. Overall, compared to ALE, RIE is faster, it cannot offer high selectivity, control ability, sharpness and especially atomic-level precision like ALE. The detailed comparison information will be illustrated in Table 3.1.

Similar to ALE, molecular layer etching (MLE) is almost similar technique. However, this term is typically used for organic or metal–organic materials, rather than inorganic materials as in ALE [35]. In his research, Young et al. [35] focused on etching metalcone thin films (particularly alucone) without the use of halogens. He proposes a 2-steps MLE cycle where the first step is dosing LOS, such as LiOtBu or LiHMDS onto the alucone surface, which cleaves bonds and creates reactive surface intermediates. Next, in step 2, trimethylaluminum (TMA) is dosed and reacts with these intermediates to form volatile complexes containing Al and organic ligands. These volatile products then desorb from the surface, effectively removing material layer by layer. The results of MLE method illustrate the same benefits as ALE: self-limiting behavior, high material selectivity, isotropic etching, chemical cleanliness, and minimal surface damage. Additionally, etching rates can be adjusted by changing the process temperature [35]. These advantages make this method suitable for thin films and delicate MEMS materials.

Plasma-free metal-organic ALE or thermal ALE is crucial for designing BioMEMS and NanoMEMS due to its atomic-level accuracy and minimal surface damage. It provides an isotropic, low-energy substitute for plasma-based etching in BioMEMS, where structural complexity and biocompatibility are crucial. This allows for the creation of accurate microfluidic channels and biosensing cavities with minimal material damage and superior surface quality [37]. Thermal ALE also enables the highly even and precise removal of material at the atomic scale, which is essential for fulfilling the strict dimensional control needed in nanoscale systems [37]. This accuracy is particularly important when developing ultra-sensitive biosensors with certain surface chemistries that need to hold up after etching [38].

Metal organic ALE enables the selective removal of metal oxides and nitrides using surface-limited fluorination and ligand-exchange reactions, allowing complicated multi-material systems to be integrated without endangering nearby layers [37, 38]. When creating layered sensor architectures and revealing functional areas in ultra-thin film sensors, this material selectivity is especially helpful [38]. Its non-damaging, low-temperature chemistry and compatibility with sensitive materials make it an essential tool in the fabrication of next-generation NanoMEMS and ultra-thin sensor platforms as device designs strive for greater density and multifunctionality [37].

Regarding industrial adoption, some of the main reasons this method has not yet been widely implemented include its slow processing speed, its reliance on a completely different technology that requires specialized tools and workflows, and the use of unstable and sensitive chemicals that make handling and integration more complex.

In terms of speed, while plasma dry etching can achieve rates of 10–500 nm per minute [31], plasma-free metal-organic ALE removes material at a much slower rate of 0.05–1 nm per cycle [39]. As a result, it may take hundreds to thousands of cycles to remove the same amount of material potentially requiring 30 minutes to an hour or more. In high-volume manufacturing, where throughput is critical, this slow rate makes ALE less suitable, particularly for bulk etching applications.

The metal-organic ALE process requires a different reactor design compared to typical dry etching systems. The reactor must offer precise control over temperature, precursor exposure, purge timing, and surface reactions. This is because the process is cyclical and self-limiting, requiring the system to:

1. Precisely feed the metal-organic precursor to ensure uniform film quality
2. Purge the chamber to remove excess and by-products
3. Introduce the second reactant, and
4. Purge again

All while ensuring that each step is fully isolated from the others to prevent unintended gas-phase reactions or contamination. This level of control is not supported by conventional plasma etching tools, making dedicated ALE reactors necessary.

However, current industrial etching tools are not designed for this level of process control. As a result, significant investments in new hardware and process development are needed to adapt this method for commercial use [36].

REFERENCES

1. Khan, F. A., Zhou, L., Kumar, V., Adesida, I., & Okojie, R. (2002). High rate etching of AlN using $BCl_3/Cl_2/Ar$ inductively coupled plasma. *Materials Science and Engineering: B*, 95, 51–54.
2. Tian, W.-C., Weigold, J. W., & Pang, S. W. (2000). Comparison of Cl_2 and f-based dry etching for high aspect ratio Si microstructures etched with an inductively coupled plasma source. *Journal of Vacuum Science & Technology B*, 18(4), 1890–1896.
3. Yang, J., Si, C., Han, G., Zhang, M., Ma, L., Zhao, Y., & Ning, J. (2015). Researching the aluminum nitride etching process for application in MEMS resonators. *Micromachines*, 6(2), 281–290.
4. Choi, S. H., Osborn, J. V., & Morgan, B. A. (2004). Chlorine-based reactive ion etching process to pattern platinum for MEMS applications. In *Materials Research Society Symposium Proceedings*, 812, F8.2.
5. Lee, S., & Kuo, Y. (2002). A new hydrogen chloride plasma-based copper etching process. *Japanese Journal of Applied Physics*, 41, 7345–7352.
6. Kim, S., Klimecky, P., Jeffries, J. B., Terry, F. L., & Hanson, R. K. (2003). In situ measurements of HCl during plasma etching of poly-silicon using a diode laser absorption sensor. *Measurement Science and Technology*, 14, 1662–1670.
7. Paul, A. K., Dimri, A. K., & Bajpai, R. P. (2002). Plasma etch models based on different plasma chemistry for micro-electro-mechanical-systems application. *Vacuum*, 68, 191–196.
8. Petit-Etienne, C., Darnon, M., Vallier, L., Pargon, E., Cunge, G., Fouchier, M., Bodart, P., Haass, M., Brihoum, M., Joubert, O., Banna, S., & Lill, T. (2011). Etching mechanisms of thin SiO_2 exposed to Cl_2 plasma. *Journal of Vacuum Science & Technology B*, 29(5), 051202.
9. Racka-Szmidt, K., Stonio, B., Żelazko, J., Filipiak, M., & Sochacki, M. (2022). A review: Inductively coupled plasma reactive ion etching of silicon carbide. *Materials*, 15(1), 123.
10. Becker, H., & Gärtner, C. (2000). Polymer microfabrication methods for microfluidic analytical applications. *Electrophoresis*, 21, 12–26.
11. Beheshti, M., Mony, S., Ebrahimi, N., & Brown, T. (2023). Reduction in scattered particles contamination in inductively coupled plasma etching systems for high volume high yield production.
12. Lee, H.-R., Jung, E.-S., Yoo, J.-U., Choi, T.-M., & Pyo, S.-G. (2024). Etching chemistry process optimization of ethylene diluted with helium (C_2H_4/He) in interconnect integration. *Micromachines*, 15(12), 1439.
13. Yoo, K. J., Chi, S. H., Yim, M. H., Han, S. B., Choi, S. J., & Ha, J. H. (2000). Dependence of metal sheet resistance on metal etch/post-etch treatment and subsequent process conditions. *Vacuum*, 59(2), 693–700.
14. Washington State Department of Health. (2010). *Chlorine Safety Guidelines: Handling, Storage, and Emergency Response*. Publication No. 331–364.
15. Lieberman, M. A., & Lichtenberg, A. J. (2005). *Principles of plasma discharges and materials processing*. Wiley, New York.
16. Vitale, S. A., Chae, H., & Sawin, H. H. (2001). Silicon etching yields in F_2, Cl_2, Br_2, and HBr high density plasmas. *Journal of Vacuum Science & Technology*, 19, 2197–2206.
17. Jin, W., Vitale, S. A., & Sawin, H. H. (2002). Plasma–surface kinetics and simulation of feature profile evolution in Cl_2 + HBr etching of polysilicon. *Journal of Vacuum Science & Technology A*, 20(6), 2106–2114.
18. Rangelow, I. W. (2001). Dry etching-based silicon micro-machining for MEMS. *Vacuum*, 62(2–3), 279–291.
19. Laermer, F., & Urban, A. (2010). Bosch-DRIE shaping MEMS – History, applications and future directions. In *Proceedings of solid-state sensors, actuators, and microsystems workshop*, Hilton Head, SC, USA, pp. 1–4.

20. Efremov, A. M., Murin, D. B., Betelin, V. B., & Kwon, K. H. (2020). Special aspects of the kinetics of reactive ion etching of SiO_2 in fluorine-, chlorine-, and bromine-containing plasma. *Russian Microelectronics*, 49, 94–102.

21. Pearton, S. J., & Norton, D. P. (2005). Dry etching of electronic oxides, polymers, and semiconductors. *Plasma Processes and Polymers*, 2(1), 16–37.

22. Pearton, S. J., & Ren, F. (1994). Science of dry etching of III–V materials. *Journal of Materials Science: Materials in Electronics*, 5(1), 1–12. https://doi.org/10.1007/BF00717875

23. Saleem, M. M., & Nawaz, H. (2019). A systematic review of reliability issues in RF-MEMS switches. *Micro and Nanosystems*, 11(1), 11–33.

24. Patrikar, R. M. (2004). Modeling and simulation of surface roughness. *Applied Surface Science*, 228(1–4), 213–220.

25. Zhao, Y. P., et al. (1999). Surface-roughness effect on capacitance and leakage current of an insulating film. *Physical Review B*, 60(12), 9157.

26. Jung, J., & Kim, K. (2023). Atomic layer etching using a novel radical generation module. *Materials*, 16(11), 3957.

27. Verhaverbeke, S., & Christenson, K. (2022). *Contamination-free manufacturing for semiconductors and other precision products* (pp. 319–330). Springer.

28. Tanaka, K., Sugiyama, T., Hamada, K., Watanabe, Y., & Mizuno, T. (2018). Bonding-based wafer-level vacuum packaging using atomic hydrogen pre-treated Cu bonding frames. *Micromachines*, 9(4), 181.

29. Oehrlein, G. S., Brandstadter, S. M., Bruce, R. L., Chang, J. P., DeMott, J. C., Donnelly, V. M., Dussart, R., Fischer, A., Gottscho, R. A, Hamaguchi, S., et al. (2024). Future of plasma etching for microelectronics: Challenges and Opportunities. *Journal of Vacuum Science & Technology B*, 42, 041501.

30. Lill, T., Vahedi, V., & Gottscho, R. (2019). *Etching of semiconductor devices*. Materials Science and Technology, 1-25

31. Kanarik, K. J., Lill, T., Hudson, E. A., Sriraman, S., Tan, S., Gottscho, R. A. (2015). Overview of Atomic layer etching in semiconductor industry. *Journal of Vacuum Science & Technology A, 33, 020802.*

32. Zhang, X., Jiang, L., & He, Y. (2018). CFD simulation and design optimization of gas flow uniformity in a dry etch chamber showerhead. *Plasma Science and Technology*, 20, 104003.

33. Schwartz, G. C. (2006). Wafer clamping and thermal management. In *Handbook of semiconductor interconnection technology* (2nd Ed.). CRC Press. https://books.google.com/books?hl=en&id=-kbMBQAAQBAJ&pg=PP6

34. Oehrlein, G., Metzler, D., & Li, C. (2015). Atomic layer etching at the tipping point: An overview. *ECS Journal of Solid State Science and Technology*, 4(6), N5041–N5053.

35. Young, M. J., Choudhury, D., Letourneau, S., Mane, A., Yanguas-Gil, A., & Elam, J. W. (2020). Molecular layer etching of metalcone films using lithium organic salts and trimethylaluminum. *Chemistry of Materials*, 32(3), 992–1001.

36. Kanarik, K. J., Lill, T., Hudson, E. A., Sriraman, S., Tan, S., Gottscho, R. A., & Sherwood, M. (2015). Atomic layer etching: Rethinking the art of etch. *Journal of Vacuum Science & Technology A*, 33, 020802.

37. George, S. M. (2020). Mechanisms of thermal atomic layer etching. *Accounts of Chemical Research*, 53(6), 1151–1160.

38. Li, C., et al. (2024). Recent advances in plasma etching for micro and nanofabrication of silicon-based materials: A review. *Journal of Materials Chemistry C*, 12, 18211–18237.

39. Sharma, V., Elliott, S. D., Blomberg, T., Haukka, S., Givens, M. E., Tuominen, M., & Ritala, M. (2021). Thermal atomic layer etching of aluminum oxide (Al2O3) using sequential exposures of niobium pentafluoride (NbF5) and carbon tetrachloride (CCl4): A combined experimental and density functional theory study of the etch mechanism. *Chemistry of Materials*, 33, 2883–2893.

4 Advanced Dry Etching Technologies for MEMS

4.1 ION BEAM ETCHING FOR HIGH-PRECISION MEMS FABRICATION

In previous chapters, we explored fluorine-free dry etching techniques commonly utilized in micro-electro-mechanical systems (MEMS) fabrication. In this chapter we delve into advanced dry etching technologies for MEMS devices. We will examine advanced approaches such as ion beam etching (IBE), laser-based etching, and hybrid etching techniques, highlighting their principles, advantages, and application scenarios. Finally, we discuss the emerging AI-driven dry etching strategies designed to enhance etching uniformity, streamline process automation, and reduce waste.

When high-energy ions or neutral particles bombard the surface of a solid material, their powerful collisions transfer enough momentum to eject atoms or atomic clusters from the surface, causing them to transition into the gas phase. This dynamic and precise method of material removal is known as sputter-etching.

IBE is a specialized variant of sputter-etching, in which material removal is accomplished by directing energized ions or neutral particles onto the surface of a solid. Unlike conventional sputter-etching, IBE distinctly separates the ion-generation region from the etching zone within the reactor. This spatial arrangement provides greater flexibility, enabling independent tuning of plasma production, ion extraction, and acceleration parameters. Additionally, IBE operates at significantly lower reactor pressures compared to traditional sputter-etching methods, supporting a broad spectrum of particle energies and densities [1].

IBE has demonstrated exceptional precision and resolution in microsystem fabrication. As stated in [2], it highlights that the achievable resolution in IBE processes is not inherently restricted by the etching method itself; instead, it primarily depends on the quality and precision of the photolithographic techniques employed to define the patterns. Consequently, advancements in photolithography directly translate into enhanced resolution in structures fabricated via IBE. Because of this high precision, IBE is especially suitable for fabricating nanoscale MEMS devices. This makes it possible to accurately produce extremely small and intricate patterns, positioning IBE as an ideal method for applications where precision at the nanoscale is crucial.

IBE provides directional etching because it uses accelerated ions directed perpendicularly toward the substrate, transferring momentum primarily along one axis. This targeted ion bombardment results in vertical sidewalls with minimal lateral etching, making IBE ideal for precise, high-resolution patterning of MEMS and nanoscale structures [1]. Directional etching is defined as a selective removal process [3] that predominantly etches vertically rather than horizontally, enabling precise control of structure dimensions and sidewall angles.

DOI: 10.1201/9781003674795-4

As IBE relies purely on physical sputtering by ion bombardment rather than chemical reactions involving fluorine-based gases, it typically provides fluorine-free. In fact, using fluorine-based gases in dry etching processes offers advantages such as increased selectivity and improved etching control, especially for SiO_2 etching [4]. However, these gases also pose significant disadvantages, including high costs and toxicity concerns. Fluorocarbon gases are environmentally harmful, contributing to ozone depletion and greenhouse effects, and pose potential health risks. Therefore, fluorine-free approaches have emerged as safer and environmentally friendly alternatives, while at the same time providing highly selective and vertical etching without the drawbacks associated with fluorine compounds.

The concept of IBE is smart and straightforward. We can summarize the whole process as "precise material removal using directional ion bombardment." In this technique, energetic ions, typically generated from an inert gas such as argon (Ar), are accelerated toward a target material, causing atoms or groups of atoms to be ejected in a process known as sputtering. If we have a patterned mask applied, ions selectively remove material from only the exposed regions, resulting in structures with clear edges and sharp vertical sidewalls [2].

As illustrated in Figure 4.1, unlike chemical-based etching, IBE avoids undesired sideways removal beneath the mask, ensuring better dimensional control and accuracy. The speed of material removal primarily depends on the energy and density of the incoming ions. Ar ions are commonly preferred due to their availability, low cost, and inertness. However, alternative ions like Xenon [5], krypton, or mercury can be utilized when higher etching rates or particular etching characteristics are necessary for specialized applications [2].

Due to its high accuracy and precision, IBE is particularly useful in making very small MEMS and NEMS devices. It is ideal for applications that demand nanoscale detail, such as MEMS sensors, RF MEMS, NEMS structures, and high-frequency resonators [2].

Another key use of IBE has been in manufacturing high-frequency quartz crystals. Traditionally, quartz resonators are made by cutting, polishing, and thinning quartz blanks to achieve the desired thickness and resonance frequency. Standard methods work well up to about 20 MHz, but beyond this, crystals become extremely thin, fragile, and expensive to produce. Using IBE, it is possible to produce quartz

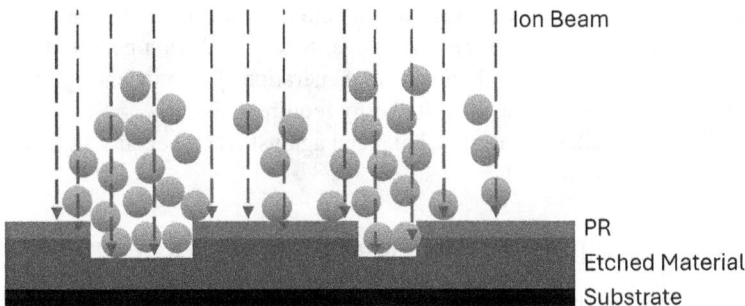

FIGURE 4.1   IBE process (self-made).

crystals for both high and low frequencies with improved precision and reliability, overcoming issues related to thinness and fragility.

Despite the advantages of precision and control, IBE has several practical limitations. Firstly, it typically exhibits a slow etching rate, as material removal depends solely on physical sputtering by energetic ions. This leads to longer processing times [6] compared to chemical or plasma etching techniques, reducing production efficiency. To address this, optimization of ion beam parameters and the use of assistive techniques such as reactive ion assistance can be employed to enhance the etch rate.

Secondly, a significant drawback is the directionality of the ion beam, which, while enhancing the etch rate and aiding in creating sharp, anisotropic features, also promotes the redeposition of sputtered material. This occurs as sputtered particles from the substrate bounce off and subsequently settle on other parts of the substrate or sidewalls of etched features. This redeposition can undermine the precision of the etching process, leading to defects and compromised pattern fidelity in microfabrication [7]. Implementing angled ion beam approaches or introducing in-situ cleaning steps during etching can help mitigate redeposition effects.

Additionally, in IBE, edge damage becomes increasingly significant as device dimensions decrease. The high-energy ions used in IBE displace atoms and create disorder zones along the edges of the etched features. This damage can alter the physical and chemical properties of the material, impacting the overall performance of the device. As the relative surface area of the edges increases in smaller devices, the effects of this damage are magnified, necessitating precise control over the ion beam to minimize imperfections and maintain device integrity [7]. Using lower energy ions and optimizing beam incidence angles can significantly reduce edge damage during the etching process.

Moreover, the IBE process demands complex and costly equipment, requiring significant investment, specialized maintenance, and skilled operators. This complexity can limit broader industrial adoption. Lastly, IBE systems have high energy consumption because substantial power is needed to generate, accelerate, and control the ion beams. This factor raises operational energy.

Moreover, the IBE process demands complex and costly equipment, requiring significant investment, specialized maintenance, and skilled operators. This complexity can limit broader industrial adoption. To alleviate these barriers, ongoing equipment design improvements aim to simplify system operation and reduce maintenance needs. Lastly, IBE systems have high energy consumption because substantial power is needed to generate, accelerate, and control the ion beams. This factor raises operational. Efforts to enhance beam generation efficiency and system design are being pursued to lower the overall energy requirements.

Overall, these challenges must be balanced against IBE's benefits when choosing it for specific fabrication applications.

4.2 LASER-BASED ETCHING FOR MEMS FABRICATION

Laser-based etching has become a vital technique for MEMS fabrication, offering precision, versatility, and non-contact processing advantages over traditional methods. This technology enables the creation of complex microstructures in diverse materials, from polymers to semiconductors, supporting applications in sensors, biomedical devices, and photonics. The following sections examine three key laser systems: excimer, femtosecond, and CO_2 lasers. Details of their working principles, MEMS applications, and technical challenges will be studied. Emerging trends and comparative analysis provide insights into selecting optimal laser etching approaches for specific MEMS requirements.

Excimer lasers operate in the ultraviolet spectrum, typically at 193 nm (ArF) or 248 nm (KrF) wavelengths, and remove material through a unique photoablation process. This mechanism involves direct photon-induced breaking of molecular bonds with minimal thermal energy transfer to the surrounding material, resulting in exceptionally precise material removal [8]. The short UV wavelength provides excellent optical resolution, enabling the fabrication of features with sub-micron precision. A notable application includes the production of inkjet printer nozzles, where excimer lasers can create precisely measured orifices with smooth sidewalls critical for controlled droplet formation. The technology can achieve etch depths of 0.1–1 μm per pulse, with the total depth precisely controlled through pulse counting.

Recent developments in excimer laser systems have introduced improved beam homogenizers and precise energy stabilization systems, enabling greater process uniformity across large substrate areas [8]. These advancements have facilitated the fabrication of microelectrode arrays with uniform feature dimensions critical for neurological sensing applications. However, excimer lasers face limitations in throughput due to relatively low pulse repetition rates (typically 100–500 Hz) and challenges in processing certain metal films that require higher ablation thresholds [8]. These throughput challenges can be mitigated through the implementation of beam-splitting techniques that enable parallel processing of multiple components simultaneously, effectively multiplying the production rate by a factor proportional to the number of beam pathways [9].

Femtosecond Lasers: Femtosecond lasers represent a significant advancement in precision material processing. They utilize ultrashort pulses, on the order of 10^{-15} seconds, to achieve material removal through multiphoton absorption and Coulomb explosion mechanisms [10]. This approach minimizes heat transfer to surrounding areas, making these lasers particularly suitable for processing thermally sensitive materials and creating high-precision features in hard substrates like silicon carbide and sapphire. Research by Liu et al. demonstrated the successful fabrication of silicon carbide pressure sensor diaphragms, where the initial surface roughness of 2.867 μm was improved to 2.143 μm through a combined laser and plasma etching process [10].

The cold ablation characteristics preserve the crystalline structure of the surrounding material, which is critical for maintaining the mechanical integrity of MEMS components. Recent advancements in femtosecond laser technology have introduced burst-mode operation, where closely spaced pulses in the nanosecond to picosecond range significantly enhance material removal rates while maintaining precision [11]. Studies have shown that optimized burst-mode parameters can increase throughput by 200–300% compared to conventional single-pulse operation, addressing one of the key limitations of this technology [11]. While offering unparalleled precision, femtosecond laser processing does face challenges in industrial adoption due to its

point-by-point processing approach, which results in relatively low throughput compared to other laser systems. Implementation of spatial light modulators and diffractive optical elements to create multiple focused spots has emerged as a promising
solution to increase processing speed without compromising precision [11].

CO_2 *Lasers:* CO_2 lasers operate at a characteristic wavelength of 10.6 μm and
function primarily through thermal ablation mechanisms. This makes them particularly suitable for processing polymers and some semiconductor materials [12]. The
technology offers several practical advantages including continuous-wave operation
that enables higher processing speeds compared to pulsed laser systems. Studies
have documented the fabrication of PMMA microfluidic channels with depths
adjustable between 18 and 660 μm and widths 110–450 μm through careful control
of laser parameters [12].

The relatively long wavelength ensures strong absorption in most polymers, while
recent advancements in beam steering systems have improved positioning accuracy
to ±5 μm, sufficient for many MEMS applications. Innovations in beam delivery systems, particularly through the development of articulated arm systems with enhanced
mirror coatings, have resulted in more uniform power distribution and improved
beam quality factor (M^2) values approaching 1.1, compared to previous generations
with values of 1.3–1.5 [13]. However, the thermal nature of CO_2 laser processing can
induce residual stresses and heat-affected zones extending 10–50 μm beyond the
processed area [12], which may affect device performance in precision applications.
These effects can be mitigated through parameter optimization and post-processing
techniques such as controlled annealing procedures that have demonstrated a 40–60%
reduction in residual stress profiles [13]. Additionally, implementing multi-pass strategies with decreasing power levels has shown promise in reducing heat-affected zones
to less than 15 μm while maintaining processing efficiency [12].

The three laser technologies each offer distinct advantages for MEMS fabrication. As indicated in Table 4.1, excimer lasers provide high precision for processing
UV-compatible materials like polymers and thin films, though they exhibit lower
throughput. Femtosecond lasers enable high-precision cold processing of materials
such as SiC, glass, and metals, but often require post-processing. CO_2 lasers offer
moderate precision solutions for materials like polymers and silicon but provide the
highest throughput.

The selection of an appropriate laser technology depends on multiple factors including material properties, required feature resolution, production volume (throughput), and economic considerations. Emerging hybrid approaches show promise by

TABLE 4.1
Comparison of Laser Technologies

Laser Type	Mechanism	Precision	Materials	Throughput	Key Challenges
Excimer	Photoablation	High (~μm)	Polymers, thin films	Low	Slow, limited metal use
Femtosecond	Cold ablation	Ultra-high	SiC, glass, metals	Moderate	Post-processing often needed
CO_2	Thermal ablation	Moderate	Polymers, silicon	High	HAZ, surface roughness

combining laser etching with conventional microfabrication techniques, such as using femtosecond lasers with reactive ion etching (RIE) to create complex 3D microstructures. These hybrid approaches not only leverage the strengths of multiple technologies but also provide effective solutions to their individual limitations [14].

APPLICATIONS OF LASER-BASED ETCHING IN MEMS

BioMEMS: Laser etching has enabled significant advancements in biomedical MEMS devices. CO_2 lasers have proven particularly effective for fabricating microfluidic networks in polymer substrates, creating smooth channel walls essential for controlled fluid flow in lab-on-chip diagnostic devices [12]. Recent developments have demonstrated the fabrication of gradient polymer scaffolds with precisely controlled pore sizes ranging from 50 to 300 μm, achieved through programmed beam parameter modulation during the etching process [15]. These structures have shown enhanced cell adhesion properties compared to conventional fabrication methods, making them promising candidates for tissue engineering applications.

Femtosecond laser processing has been successfully applied to create biocompatible silicon carbide structures for implantable sensors, where surface quality and material integrity are critical [10]. The technology's unique ability to modify material properties without thermal damage has been leveraged to create selective cell adhesion patterns on biocompatible substrates through localized surface chemistry modifications. Additionally, the technology has been used to produce microneedle arrays for transdermal drug delivery systems, demonstrating the versatility of laser-based approaches in medical applications. Excimer lasers have been employed to create high-aspect-ratio microchambers in biodegradable polymers for controlled drug release systems, where precise dimensional control is essential for achieving specific pharmacokinetic profiles [15].

Photonic MEMS: In photonic applications, femtosecond lasers have enabled the fabrication of optical waveguides through localized refractive index modification in glass substrates [16]. The technology allows for three-dimensional patterning of optical circuits with minimal insertion losses. Advanced beam shaping techniques, particularly using spatial light modulators, have further enhanced processing capabilities, enabling the creation of complex diffraction gratings with variable line spacing for specialized spectroscopic applications [16].

Excimer lasers have been employed to create diffractive optical elements with subwavelength features for miniature spectrometers and other optical MEMS devices. Recent advancements include the development of polarization-sensitive elements through precisely controlled ablation patterns that introduce form birefringence at the nanoscale [14]. The precision and flexibility of laser etching techniques have opened new possibilities in integrated photonics, enabling complex optical functionalities on chip-scale devices. CO_2 laser processing has found application in the fabrication of refractive micro-lens arrays in polymer substrates, where controlled thermal effects are harnessed to create smooth optical surfaces through localized material reflow, achieving surface roughness values below 5 nm after optimization [14].

The integration of laser-based etching with established microfabrication techniques represents a significant advancement in MEMS manufacturing capabilities.

Laser-assisted deep reactive ion etching (LA-DRIE) combines the precision of laser processing with the high aspect ratio capabilities of DRIE, enabling the creation of complex 3D structures that would be challenging to achieve with either technology alone [17]. This hybrid approach has demonstrated promise for the fabrication of through-silicon vias (TSVs) in 3D-integrated MEMS packages, where laser pre-patterning significantly improves etch anisotropy and reduces processing time compared to conventional DRIE alone.

Complementary approaches involving laser pre-patterning followed by wet chemical etching have shown remarkable results in crystalline materials, where laser-induced stress fields can influence subsequent etch selectivity. This technique has been successfully applied to create high-aspect-ratio structures in quartz substrates for resonator applications, achieving etch ratios exceeding 20:1 along specific crystallographic planes [17]. The synergistic implementation of multiple etching modalities not only addresses the limitations of individual techniques but also creates new manufacturing possibilities that expand the design space for MEMS devices.

Surface quality remains a critical consideration in laser-based MEMS fabrication. The ablation process can induce micro-scale surface textures through material redeposition and thermal effects, particularly in thermal ablation processes like those using CO_2 lasers [12]. For femtosecond laser processing of silicon carbide, studies have shown that the initial surface roughness of 2.867 μm can be improved to 2.143 μm through additional plasma etching steps [10].

Various post-processing techniques including chemical polishing, plasma treatment, and laser remelting have been developed to address surface roughness issues and meet the stringent requirements of optical and fluidic applications. Recent innovations in laser parameter optimization have demonstrated significant improvements in as-processed surface quality. For instance, the implementation of temporally shaped pulse trains with optimized energy distribution has shown a 30–40% reduction in surface roughness compared to conventional single-pulse processing [11]. Additionally, hybrid approaches combining laser ablation with localized electrochemical polishing have achieved sub-10 nm surface roughness in metallic components, making them suitable for high-performance MEMS applications such as micro-mirrors and precision actuators [11].

Thermal effects present a significant challenge in laser processing, particularly for CO_2 lasers and to a lesser extent for femtosecond systems. The thermal gradients created during processing can alter material properties near the etched features, potentially affecting device performance and reliability [12]. Various mitigation strategies have been developed, including beam defocusing techniques, protective coatings, and optimized scanning patterns. These help protect areas not to be etched.

Advanced thermal modeling approaches have also been employed to predict and minimize these effects during the design phase. Recent developments in process optimization have introduced adaptive cooling systems that provide localized temperature regulation during processing, reducing heat-affected zone extent by up to 65% in polymer substrates [13]. Innovative scanning strategies, particularly those implementing variable dwell times and interlaced patterns, have demonstrated significant reductions in thermal accumulation effects while maintaining processing throughput. For metals and semiconductors, the application of specialized coatings with high thermal conductivity has proven effective in dissipating heat away from the processing zone, minimizing structural and compositional changes in the surrounding material [13].

Material ejection and subsequent redeposition present significant challenges in laser-based MEMS fabrication, particularly for high-aspect-ratio structures, where ejected debris can compromise dimensional accuracy and surface quality. Traditional vacuum-based extraction systems have shown limited effectiveness for microscale features due to complex flow dynamics in confined spaces. Recent innovations have introduced localized gas jet systems that create controlled flow patterns precisely tailored to feature geometry, improving debris removal efficiency by up to 80% compared to conventional approaches [9]. For applications requiring exceptional cleanliness, in-situ plasma cleaning protocols have been developed to remove residual contaminants immediately following laser processing, preventing material rebonding and preserving feature integrity. Additionally, specialized immersion processing techniques, where ablation occurs within a fluid medium, have demonstrated promising results for controlling debris dispersion and facilitating immediate removal of ablated material [9].

Future Trends and Outlook: The field of laser-based MEMS fabrication continues to evolve rapidly, with several promising developments on the horizon. Burst-mode femtosecond lasers are showing potential for significantly improving processing speeds while maintaining precision. Hybrid approaches that combine different laser types or integrate laser processing with other microfabrication techniques are enabling new capabilities in device manufacturing [12, 16].

Additionally, advancements in beam shaping and control systems are providing greater flexibility in pattern generation and feature definition. The integration of machine learning (ML) algorithms for process parameter optimization has demonstrated significant improvements in both processing speed and quality outcomes. These adaptive systems analyze real-time process data to dynamically adjust parameters such as pulse energy, repetition rate, and scanning strategy, reducing optimization time by up to 75% compared to conventional iterative approaches [11]. Looking forward, the development of compact fiber-based femtosecond laser systems promises to dramatically reduce equipment footprint and cost, potentially broadening the accessibility of high-precision laser processing for MEMS applications [11].

Environmental considerations are increasingly influencing the development of laser-based manufacturing technologies. Dry laser processing offers significant advantages over conventional wet etching approaches that typically generate substantial chemical waste. Research indicates that laser-based fabrication can reduce process-related chemical consumption by 60–80% for comparable components [17]. The implementation of closed-loop material reclamation systems, particularly for precious metals in sensor applications, has demonstrated both economic and environmental benefits. These systems capture and recycle ablated material, reducing raw material requirements while minimizing waste generation. Additionally, the lower energy consumption of modern fiber laser systems, which can offer wall-plug efficiencies exceeding 30% compared to 10–15% for traditional systems, contributes to reduced carbon footprints for manufacturing operations [17]. These sustainability advantages are expected to drive further adoption of laser-based approaches as environmental regulations become increasingly stringent.

The versatility and precision of laser-based etching technologies continue to enable new applications across diverse fields. In energy harvesting, laser-structured surfaces with optimized micro- and nano-topographies have demonstrated enhanced performance in triboelectric generators, with output improvements of 30–40% compared to conventional designs [15]. The aerospace sector has embraced laser etching for fabricating microscale features in specialized materials such as ceramic matrix composites and high-temperature alloys, where conventional machining approaches face significant limitations. These applications leverage the material-agnostic nature of laser processing to create precision components for harsh operating environments. Additionally, the quantum computing field has begun to explore femtosecond laser fabrication for creating ion traps and waveguide arrays with nanoscale precision, opening new possibilities for scalable quantum device architectures [15]. As these emerging applications mature, they will likely drive further innovations in laser technology specifically tailored to their unique requirements.

4.3   HYBRID DRY ETCHING TECHNIQUES FOR MEMS

Hybrid etching methods combine different techniques to improve process efficiency and material selectivity. Microfabrication is essential for the progression of semiconductor and MEMS technologies, with etching techniques being pivotal for the attainment of high-precision structures. Therefore, diverse hybrid etching techniques, encompassing RIE, inductively coupled plasma reactive ion etching (ICP-RIE), and direct laser writing (DLW) in conjunction with wet etching are discussed. These approaches are crucial for the fabrication of microstructures, including thin-film bulk acoustic wave (BAW) resonators and SiO_2 microcantilevers, where precise etch rates, anisotropy, and homogeneity are critical. The incorporation of AI-driven process management in dry etching improves fabrication precision, optimizing etch profiles and reducing errors. Utilizing these sophisticated processes, microfabrication facilitates next-generation applications in electronics, sensing, and MEMS development.

Reactive IBE: RIE processes must be optimized for precision and efficiency in microfabrication. A key factor is the etch rate, which defines the depth of material removed per unit time. Higher etch rates improve wafer throughput and reduce costs but must remain controllable. In high-aspect-ratio features, the etch rate often decreases with depth due to restricted diffusion of etchant gases and difficulty in removing reaction by-products. Another critical aspect is etch uniformity, where the ideal process maintains a consistent etch rate across the wafer. However, variations commonly occur due to non-uniform plasma distribution and gas flow, often resulting in a "bull's-eye" pattern with concentric etch rate variations from the wafer's center to its edge. Optimizing plasma power, gas composition, and pressure can improve uniformity. By carefully managing these parameters, RIE ensures precise and reproducible etching, making it indispensable for semiconductors and MEMS applications [18].

In the fabrication of thin-film BAW resonators, the etching of the aluminum nitride (AlN) piezoelectric layer is a critical process that directly impacts the device's electro-acoustic properties, including insertion loss, electromechanical coupling, and quality factor. The etch method must provide a high etch rate, good anisotropy, and precise sidewall angle control. AlN can be etched using chlorine (Cl)-based plasma, which forms volatile $AlCl_3$ at temperatures above 180°C or Al_2Cl_6 at room temperature. Since this process is predominantly chemical, the etch profile tends to be isotropic. To enhance anisotropy, additional chemical species can be introduced to protect the sidewalls. However, Cl-based plasma etching has disadvantages, such as the corrosive nature of both the process gases and reaction by-products. An alternative approach utilizes fluorine (F)-based plasmas, but etching AlN with F alone is inefficient because it forms AlF_3, a non-volatile compound.

$$3AlN + 12F \rightarrow 3AlF_3 + N_2 + NF_3 \qquad (4.1)$$

To overcome this, Ar is introduced, and bias voltage is increased, allowing ion bombardment to remove AlF_3 and sustain the etch process [19]. Studies using RIE with an SF_6/Ar plasma mixture (40% SF_6, 1 kV bias, and 15°C substrate temperature) demonstrated an etch rate of approximately 135 nm/min, with vertical sidewalls and no damage to the tungsten (W) bottom electrode. Despite the challenges of using Cl-based gases such as Cl_2 and BCl_3, they remain widely used in RIE and ICP-RIE processes. Experiments with BCl_3/Ar discharges showed etch rates of 10–20 nm/min, increasing with DC bias and pressure. The following chemical reaction steps results during Cl_2/BCl_3-based plasma. Firstly, the absorption of chlorine gases to the AlN surface takes place via a chemical reaction.

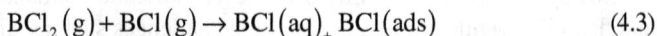

$$Cl(g) \rightarrow Cl(ads) \qquad (4.2)$$

$$BCl_2(g) + BCl(g) \rightarrow BCl(aq)_+ BCl(ads) \qquad (4.3)$$

The dissolved molecules of BCl_2 dissociate and results in a highly reactive Cl species at the surface:

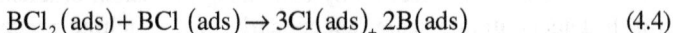

$$BCl_2(ads) + BCl(ads) \rightarrow 3Cl(ads)_+ 2B(ads) \qquad (4.4)$$

The AlN film reacts with the adsorbed Cl species, giving out the product:

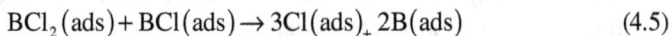

$$BCl_2(ads) + BCl(ads) \rightarrow 3Cl(ads)_+ 2B(ads) \qquad (4.5)$$

And finally, the product and residues desorb from the surface:

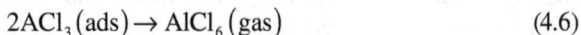

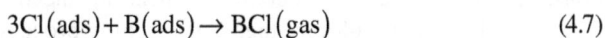

$$2ACl_3(ads) \rightarrow AlCl_6(gas) \qquad (4.6)$$

$$3Cl(ads) + B(ads) \rightarrow BCl(gas) \qquad (4.7)$$

In CH_4/H_2/Ar plasmas, AlN etching was largely dominated by physical-ion bombardment. Although no single condition favored AlN etching over indium nitride (InN) or gallium nitride (GaN), proper process optimization could achieve selectivity, ensuring effective material removal without damaging underlying layers [20].

Direct Laser Etching and Wet Etching for micro-patterning: The fabrication of SiO$_2$ microcantilevers (MCs) using a combination of DLW and wet etching techniques offers a significant advancement in micro/nano fabrication. This approach combines the precision of laser-based patterning with the efficiency of wet etching to achieve highly controlled microstructures [21]. DLW is a mask-free fabrication method that enables rapid prototyping and design modifications without requiring photomasks. Its ability to achieve a minimum feature size of 150 nm makes it an ideal technique bridging the gap between nanolithography and micrometer-resolution processes. The fabrication process begins with wafer cleaning, followed by spin-coating a photoresist layer. After pre-baking, the optimized laser dose transfers the desired pattern onto the wafer, which then undergoes development and post-baking.

Following DLW, wet etching is employed to release the SiO$_2$ MCs. This is achieved using a TMAHw solution that selectively removes the bulk silicon, freeing the MCs from the substrate. One of the major challenges in wet etching is "stiction," where MCs adhere to the substrate due to capillary forces. To address this, a sharp convex tip is introduced at the free end of the MCs, reducing the contact area and facilitating detachment. Additionally, post-etch rinsing in boiling deionized (DI) water minimizes surface tension and eliminates trapped air bubbles, improving the overall yield of released MCs. To ensure precision and effectiveness, key fabrication parameters were optimized, enabling successful pattern transfer and MC release. The resulting MCs were characterized using a 3D optical microscope and a nano-vibration analyzer to evaluate their mechanical properties and resonance frequencies, which are crucial for sensing applications [22].

This hybrid approach to microcantilever fabrication enhances process efficiency and flexibility while addressing common challenges such as stiction. By combining DLW and wet etching, this method demonstrates significant potential for advancing micro/nano fabrication technologies, particularly in sensor applications.

AI-driven hybrid dry etching control for MEMS: AI significantly improves dry etching in MEMS fabrication by enhancing precision, efficiency, and defect control. It detects atomic-scale defects early in the etching process, reducing wafer rejections and improving overall yield. Predictive modeling enables precise control over the etch profile by analyzing plasma parameter variations, which is crucial for fabricating sub-3nm nodes and high-k dielectric layers. Reinforcement learning allows successful etching strategies to be adapted across different fabrication facilities, considering local constraints such as material availability and regulatory requirements. Collaborative multi-agent systems facilitate real-time data sharing among fabs, improving consistency and accelerating problem resolution. Additionally, AI-powered digital twins provide real-time monitoring and simulations, optimizing etching outcomes to meet advanced semiconductor standards. By integrating AI at multiple stages, dry etching processes become more reliable, efficient, and adaptable, driving MEMS manufacturing advancements [23]. AI, particularly through ML techniques like long short-term memory (LSTM) networks, plays a crucial role in improving the accuracy of endpoint detection in dry etching processes. Traditional methods often rely on specific signal patterns, which can lead to inaccuracies. In contrast, AI models can analyze the entire time series data collected during the etching process, leading to more reliable endpoint detection [24]. Also, by Utilizing Time Series Data, the optical emission spectroscopy (OES) time series data contains valuable information throughout the etching process. AI algorithms can effectively leverage this data by considering long-term temporal dependencies, which traditional methods often overlook. This comprehensive analysis allows for better decision-making and reduces the false detection rate [25].

ICP-RIE: ICP-RIE is a sophisticated etching technique that combines the benefits of both inductively coupled plasma and RIE, allowing for high-density plasma generation and enhanced etching precision. This method is particularly effective for etching materials like AlN and other III-N group semiconductors [23, 27].

The ICP-RIE process operates by utilizing a coil to generate high-density plasma, which is essential for achieving anisotropic etching profiles. The independent control of plasma density and ion energy is facilitated by varying the ICP power and substrate bias, leading to improved etching characteristics. One of the key advantages of ICP-RIE is its ability to achieve high etch rates while maintaining excellent selectivity over different materials. For instance, in the etching of AlN, optimized conditions can yield etch rates exceeding 177 nm/min with a selectivity ratio of 67:1 over nickel (Ni) [23].

The etching process is influenced by various parameters, including gas flow rates, chamber pressure, and bias voltage. Adjusting these parameters can significantly affect etch rate and roughness surface, making meticulous control essential for achieving desired outcomes [27]. ICP-RIE is widely used in semiconductor manufacturing for patterning intricate micro- and nano-scale features, which are critical for the development of advanced electronic devices (see Figure 4.2). Its application spans across various sectors, including MEMS, optoelectronics, and RF filters. The technique also faces challenges, such as micro-trenching effects, where etch rates may vary near the edges of features, leading to potential defects. Understanding and mitigating these effects is crucial for improving the overall quality of the etched structures. Overall, ICP-RIE represents a significant advancement in etching technology, enabling the fabrication of high-performance devices with complex geometries while addressing the challenges associated with material compatibility and process optimization. Figure 4.2 shows a scheme for an ICP-RIE reactor, in which the cathode is powered by a 13.56 MHz RF generator, and has a smaller surface area compared to the larger anode (lower electrode). A coil within the reactor generates a magnetic field that confines the plasma, minimizing electron scattering against the chamber walls. The material being etched is placed on a holder connected to a second RF generator, which attracts and accelerates ions from the plasma. The etched material has a negative plasma potential that can reach several hundred volts. The reactor allows for independent control of the RF power from both generators. The inductive power (PICP) influences the plasma flux, while the RF bias power (PRIE) is responsible for accelerating plasma ions and creating the polarization voltage on the etched material. Both power sources significantly affect the etching rate [26].

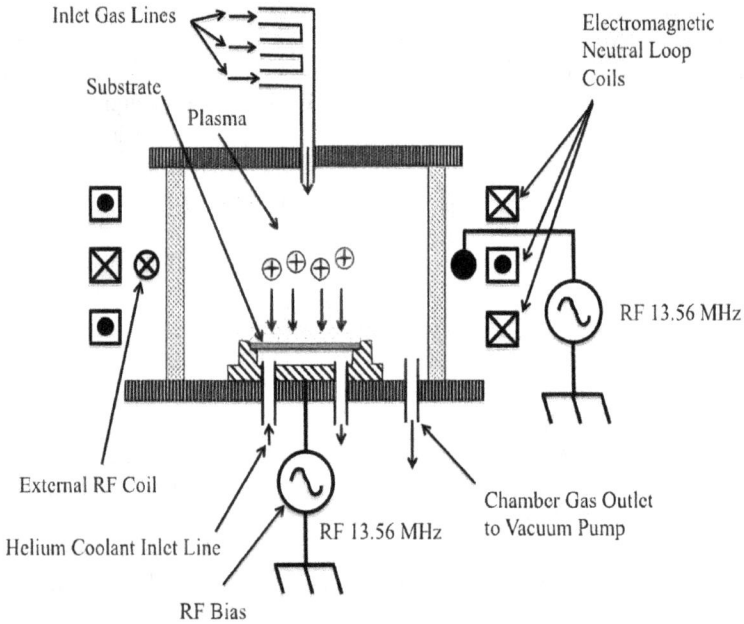

FIGURE 4.2   Scheme of an ICP-RIE reactor [26].

4.4 AI-DRIVEN DRY ETCHING OPTIMIZATION

In the evolution of MEMS fabrication, advanced etching techniques such as IBE, laser beam etching, and hybrid methods have played a pivotal role in achieving the precision and miniaturization required for modern devices. IBE offers excellent directionality but suffers from surface damage and limited material selectivity. Laser beam etching provides high spatial resolution but can introduce thermal stress and roughness edge. While hybrid techniques aim to merge the benefits of multiple methods, they often add complexity and increase process variability.

To overcome these challenges, artificial intelligence (AI) and ML are being increasingly integrated into dry etching workflows. These technologies offer real-time monitoring, adaptive process control, and predictive capabilities that reduce trial-and-error cycles. By learning from process parameters and historical outcomes, AI systems can optimize gas ratios, chamber conditions, and power levels to achieve desired etch profiles with greater consistency.

Beyond improving performance, these intelligent systems contribute to the goals of green microfabrication, minimizing material waste, lowering energy consumption, and enabling more sustainable process development. As the demands for high-yield, high-precision MEMS devices grow, AI-driven etching optimization is poised to become a cornerstone of next-generation fabrication strategies. The following sections explore this shift in detail, focusing on plasma etch control, real-time defect monitoring, and pattern recognition within the fabrication process.

ML-based plasma etch control: Plasma-based dry etching is a cornerstone of MEMS fabrication, offering high anisotropy, selectivity, and precision. However, its effectiveness is limited by the complex interdependence of parameters such as gas flow rates, chamber pressure, RF power, and temperature. Traditionally, engineers rely on trial-and-error or empirical recipe tuning, which can be time-consuming and resource-intensive. ML offers a data-driven alternative that can model nonlinear plasma behavior, optimize recipes, and enable adaptive process control.

Supervised learning methods, including support vector machines (SVMs), random forests, and artificial neural networks (ANNs), have been used to predict critical outcomes like etch rate, uniformity, and sidewall angle. For example, models trained on CF_4/O_2 flow ratios, power settings, and pressure, with some achieving R^2 values above 0.98 for etch depth and dimensional accuracy. In another case, regression trees were used to control sidewall angles in high-aspect-ratio etching, enabling tighter tolerance control with minimal iterations [28].

More recent techniques, such as deep reinforcement learning using Deep Q-Networks (DQN), have been applied to optimize deep reactive ion etching (DRIE) profiles by dynamically adjusting passivation-etch cycle timing to achieve vertical sidewalls with improved surface quality [29]. Additionally, Gaussian Process Regression (GPR) has shown remarkable accuracy in predicting etch rate and sidewall angle for ScAlN films, even when trained on limited experimental datasets, highlighting its efficiency for recipe optimization in emerging MEMS materials [30]. Recent advances also include deep learning models such as attention-enhanced LSTMs for endpoint detection using OES time-series data. The model achieved 98.2% accuracy in detecting etch completion, outperforming traditional classifiers [24]. Similarly, transformer-based models have been developed for atomic layer etching (ALE) endpoint control and integrated with run-to-run feedback systems to dynamically tune etch parameters based on prior batches [31].

Moreover, green microfabrication goals are supported through waste reduction and efficiency gains. ML can even be applied to bio-based microfabrication, predicting device performance from low-data process sets. Across these examples, AI contributes to smarter, more sustainable dry etching by minimizing rework, reducing power consumption, and accelerating process development [32].

Real-time defect monitoring in MEMS fabrication: Defect monitoring during MEMS fabrication is crucial, as even sub-micron imperfections can significantly degrade device reliability and performance. Traditional inspection techniques such as scanning electron microscopy (SEM) are accurate but limited by their offline nature and manual interpretation. This creates delays in defect identification, often after the wafer has progressed beyond the corrective stage. The integration of AI, particularly deep learning, enables in-line, real-time defect detection with high precision and responsiveness.

Convolutional neural networks (CNNs) are at the forefront of this shift. Architectures like faster R-CNN and adaptive deep CNNs have been applied to SEM images for identifying surface anomalies, pattern discontinuities, and etch defects [26]. These models not only classify defects with high accuracy but also localize them spatially, enabling immediate intervention. A CNN trained on plenoptic images of medical MEMS wafers achieved 98% precision in detecting multiple defect types, such as cracks and particulate contamination—highlighting its viability in sensitive, high-stakes applications [33].

Beyond classification, real-time AI systems are now integrated into feedback control loops, closing the gap between defect detection and process correction. LSTM-based endpoint models were also noted for their potential in anomaly detection, improving process robustness against unexpected variations [24].

By automating defect monitoring, these AI-driven systems reduce yield loss, minimize manual labor, and contribute to more resource-efficient, greener microfabrication—meeting both productivity and sustainability goals.

AI-driven MEMS pattern recognition: AI and ML have demonstrated growing utility in recognizing patterns both in MEMS sensor output and within fabrication process data. While post-fabrication applications such as motion tracking, inertial sensing, and user behavior interpretation are well-established, the focus is now shifting upstream toward identifying and acting on patterns during microfabrication itself.

AI models such as LSTM networks and neural sensor fusion frameworks were shown to enhance post-fabrication pattern recognition in MEMS-based navigation systems [34]. These models corrected signal drift, reduced noise, and improved real-time positioning accuracy in wearable and UAV platforms. While these applications illustrate AI's power at the device level, the same pattern-learning capabilities are increasingly relevant within the fabrication environment.

This approach is further demonstrated in the context of neuromorphic device design, where ML models have been used to predict electrical performance based on bio-organic fabrication parameters. A neural network trained on honey-based RRAM devices successfully forecasted SET voltage distributions using only limited process data, showing the potential of ML to enhance green and sustainable device fabrication through pattern learning [32]. Dry etching and thin film deposition processes generate rich, high-dimensional datasets—from SEM images to line edge roughness (LER) metrics. CNNs and deep neural networks (DNNs) trained on edge shape descriptors and image-based surface data have been used to predict etched profile outcomes and material layer behavior [35]. Such models can predict issues like profile deformation and layer nonuniformity, enabling early intervention and recipe adjustment.

Pattern recognition also supports predictive quality control by learning from fabrication process signatures. ML models trained on process morphology and etch profile trends can forecast reliability issues before device testing [28]. This evolution from passive inspection to proactive planning is pivotal for achieving higher yield, process transparency, and long-term manufacturing sustainability.

The integration of AI and ML into MEMS dry etching processes marks a pivotal advancement in both precision manufacturing and sustainable microfabrication. From optimizing plasma etch parameters to enabling real-time defect detection and predictive pattern recognition, intelligent models are redefining how fabrication is controlled, monitored, and improved. These techniques not only reduce material waste and power consumption but also shorten development cycles and enhance overall process reliability.

As the complexity and performance demands of MEMS devices continue to rise, the role of AI will only grow more central. By transforming etching from a fixed, recipe-driven procedure into an adaptive, feedback-informed process, AI paves the way for a more resilient, efficient, and scalable manufacturing pipeline.

REFERENCES

1. Köhler, M. (1999). *Etching in microsystem technology*. John Wiley & Sons.
2. Norgate, P., & Hammond, V. J. (1974). Ion beam etching. *Physics in Technology*, 5, 186–190.
3. Peng, B., Ding, S., Yao, H., & Wen, C. Y. (2021). Directional etching of silicon: From crystal orientation dependent to crystal orientation independent. *Journal of Micromechanics and Microengineering*, 31(9), 095002.
4. Abdolahad, M., Moradi, A. R., Shahhoseini, H. R., & Mohajerzadeh, S. (2010). Highly selective and vertical SiO_2 etching on silicon substrates using a fluorocarbon-free DC plasma. *Journal of Physics D: Applied Physics*, 43(39), 395402.
5. Mayr, S., Plank, S., Hofer, F., & Schaffer, B. (2021). Xenon plasma focused ion beam milling for obtaining soft X-ray transparent samples. *Crystals*, 11(5), 546.
6. Lee, R. E. (1984). Ion-beam etching (milling). In N. G. Einspruch & D. M. Brown (Eds.), *VLSI electronics: Microstructure science* (vol. 8, pp. 341–364). Elsevier.
7. Peng, X., et al. (2009). Towards the sub-50 nm magnetic device definition: Ion beam etching (IBE) vs plasma-based etching. *Vacuum*, 83(6), 1007–1013.
8. Holmes, A. S. (2001). Laser fabrication and assembly processes for MEMS. In Proceedings of SPIE (Vol. 4274).
9. Sugioka, K., & Cheng, Y. (2014). Ultrafast lasers—Reliable tools for advanced materials processing. *Light: Science & Applications*, 3, e149.
10. Liu, X. Q., Bai, B. F., Chen, Q. D., & Sun, H. B. (2019). Etching-assisted femtosecond laser modification of hard materials. *Opto-Electronic Advances*, 2, 19002101.
11. Stoian, R., Colombier, J. P., Mauclair, C., Cheng, G., Bhuyan, M. K., & Velpula, P. K. (2018). Spatial and temporal laser pulse design for material processing on ultrafast scales. *Applied Physics A*, 124, 97.
12. Vidya, S., Wattal, R., Singh, L., & Mathiyalagan, P. (2021). CO_2 laser micromachining of polymethyl methacrylate (PMMA): A review. In R. M. Singari, K. Mathiyazhagan, & H. Kumar (Eds.), *Advances in manufacturing and industrial engineering* (pp. 755–768). Springer.
13. Snakenborg, D., Klank, H., & Kutter, J. P. (2004). Microstructure fabrication with a CO_2 laser system. *Journal of Micromechanics and Microengineering*, 14(2), 182–189.
14. Gattass, R. R., & Mazur, E. (2008). Femtosecond laser micromachining in transparent materials. *Nature Photonics*, 2(4), 219–225.
15. Bellouard, Y., Champion, A., McMillen, B., Mukherjee, S., Thomson, R. R., Pépin, C., Gillet, P., & Cheng, Y. (2016). Stress-state manipulation in fused silica via femtosecond laser irradiation. *Optica*, 3(12), 1285–1293.
16. Zheng, J. X., Tian, K. S., Qi, J. Y., Guo, M. R., & Liu, X. Q. (2023). Advances in fabrication of micro-optical components by femtosecond laser with etching technology. *Optics and Laser Technology*, 167, 109793.
17. Madou, M. J. (2018). *Fundamentals of microfabrication and nanotechnology* (4th ed.). CRC Press.
18. Huff, M. (2020). *Process variations in microsystems manufacturing*. Springer. https://doi.org/10.1007/978-3-030-40560-1
19. Chen, D., Xu, D., Wang, J., Zhao, B., & Zhang, Y. (2008). Dry etching of AlN films using the plasma generated by fluoride. *Vacuum*, 83(2), 282–285.
20. Pearton, S. J., Abernathy, C. R., Ren, F., Lothian, J. R., Wisk, P. W., & Katz, A. (1993). Dry and wet etching characteristics of InN, AlN, and GaN deposited by electron cyclotron resonance metalorganic molecular beam epitaxy. *Journal of Vacuum Science & Technology B*, 11(4), 1772–1775.
21. Middya, S., & Prabakar, K. (2017). Si orientation dependent release of SiO_2 microcantilevers by wet chemical etching method. *arXiv*. https://arxiv.org/abs/1706.10126

22. Balasubramanian, S., & Prabakar, K. (2019). Fabrication and characterization of SiO$_2$ microcantilevers by direct laser writing and wet chemical etching methods for relative humidity sensing. *Microelectronic Engineering*, 212, 61–69.

23. Ramachandran, A. (2023). Revolutionizing semiconductor etching: Artificial intelligence in RIE, ALE, and high-aspect ratio features for next-generation manufacturing. *Unpublished manuscript*. Retrieved from https://www.researchgate.net/publication/386708385

24. Kim, Y. J., et al. (2024). Improved plasma etch endpoint detection using attention-based long short-term memory machine learning. *Electronics*, 13(17), 3577.

25. Chakroun, I., Ashby, T., Das, S., Halder, S., Wuyts, R., & Verachtert, W. (2020). Using unsupervised machine learning for plasma etching endpoint detection. In *Proceedings of the 9th International Conference on Pattern Recognition Applications and Methods (ICPRAM)* (pp. 273–279).

26. Huff, M. (2021). Recent advances in reactive ion etching and applications of high-aspect-ratio microfabrication. *Micromachines*, 12(8), 991.

27. Pinto, R., Gund, V., Calaza, C., Nagaraja, K. K., & Vinayakumar, K. B. (2022). Piezoelectric aluminum nitride thin-films: A review of wet and dry etching techniques. *Microelectronic Engineering*, 257, 111753.

28. Podder, I., Fischl, T., & Bub, U. (2023). Artificial intelligence applications for MEMS-based sensors and manufacturing process optimization. *Telecom*, 4(1), 165–197.

29. Wang, F., Yu, H., Miao, Y., He, Y., Sun, K., & Sun, Y. (2025). Deep reinforcement learning-based parameters optimize prediction model for smooth-vertical sidewall profile in deep reactive ion etching process. In *Proceedings of the 2025 IEEE 38th International Conference on Micro Electro Mechanical Systems (MEMS)* (pp. 565–568).

30. Higashi, K., Miyazaki, T., & Tanaka, S. (2024). Machine learning-based recipe optimization of the ScAlN etching process for MEMS applications. *ACS Applied Electronic Materials*, 6(1), 161–169.

31. Wang, H., Ou, F., Suherman, J., Orkoulas, G., & Christofides, P. D. (2025). Integration of on-line machine learning-based endpoint control and run-to-run control for an atomic layer etching process. *Digital Chemical Engineering*, 14, 100206.

32. Vicenciodelmoral, A. Y., Tanim, M. M. H., Zhao, F., & Zhao, X. (2023). A machine learning approach to support neuromorphic device design and microfabrication. In Proceedings of the 2023 International Conference on Machine Learning Applications (ICMLA) (pp. 1627–1634).

33. Amini, A., Kanfoud, J., & Gan, T. H. (2021). An artificial-intelligence-driven predictive model for surface defect detections in medical MEMS. *Sensors*, 21(18), 6141.

34. Ali, A., Saltykova, O. A., & Basokhbat, N. A. (2023). Artificial intelligence driven optimization of MEMS navigation sensors for enhanced user experience. *Vestnik Rossiiskogo Universiteta Druzhby Narodov. Seriya: Inzhenernye Issledovaniya*, 24(4), 305–322.

35. Kondi, A., Papia, E. M., & Constantoudis, V. (2024). Machine learning applications in nanotechnology manufacturing: From etching accuracy to deposition prediction. In *Proceedings of the 13th Hellenic Conference on Artificial Intelligence* (pp. 1–6).

5 Wet Etching Alternatives for Green MEMS Processing

5.1 ORGANIC ACID-BASED WET ETCHING

The rapid evolution of microelectromechanical systems (MEMS) has fueled demand for safer and more sustainable microfabrication processes. Among these, wet chemical etching remains indispensable for its simplicity, scalability, and cost-effectiveness in defining microscale patterns, structures, and features. However, conventional wet etching typically relies on hazardous chemicals, such as hydrofluoric acid (HF), nitric acid (HNO_3), and strong alkaline solutions, posing significant health and environmental risks. To mitigate these concerns, organic acid-based wet etching has emerged as a promising, environmentally friendly alternative for MEMS applications [1, 2].

Organic acids such as citric acid, oxalic acid, and lactic acid possess unique chelating and complexation properties that enable controlled material dissolution with reduced environmental impact. These acids are biodegradable, exhibit low toxicity, and are highly versatile, facilitating the etching of semiconductors, metals, oxides, glass, and quartz substrates. Their broad applicability extends to MEMS, microfluidics, optoelectronics, and bio-integrated devices [1, 2].

Among these, citric acid—often used in combination with hydrogen peroxide (H_2O_2)—has demonstrated high efficacy in etching III-V compound semiconductors such as GaAs, AlGaAs, and InP. The etching mechanism involves the formation of soluble citrate-metal complexes, enabling selective material removal while minimizing damage to underlying layers and photoresist structures. DeSalvo et al. reported that optimized citric acid-H_2O_2 formulations achieved etch selectivity ratios exceeding 400 for InGaAs/InP heterostructures, highlighting the potential of this system for optoelectronic and high-frequency MEMS applications, where precise pattern transfer and surface integrity are critical [3].

The performance of citric acid-based etching is highly sensitive to process parameters such as acid concentration, oxidant ratio, temperature, and solution agitation. Increased H_2O_2 content enhances surface oxidation and etch rates; however, excessive oxidant concentrations can lead to undesirable surface roughness, underscoring the need for precise process control [3, 4]. Low-temperature processes using citric acid have been shown to reduce lateral undercutting, enabling high-resolution feature definition for dense MEMS structures [2].

Oxalic acid has also garnered attention for applications that demand superior surface quality and dimensional precision. Its dual functionality as both an etchant and a passivating agent enables the formation of smooth, low-roughness surfaces with

DOI: 10.1201/9781003674795-5

minimal lateral etching. Lee et al. demonstrated that oxalic acid etching of Ga-doped ZnO thin films yields sharper edge profiles and smoother surfaces than conventional mineral acid-based processes, making it highly suitable for optical MEMS, transparent electrodes, and advanced display technologies [5].

Nevertheless, oxalic acid's strong chelating capacity can induce the formation of insoluble surface complexes that passivate the material, inhibiting etch rates. While this passivation effect benefits surface smoothness, it significantly slows etching, limiting throughput for high-volume MEMS production. In copper etching, oxalic acid precipitates cupric oxalate, suppressing etch rates while improving surface finish and dimensional control—attributes essential for precision MEMS structures [6].

Lactic acid, though less commonly used as a primary etchant, plays a vital role in bioMEMS and microfluidic devices. Its excellent biocompatibility and surface functionalization capabilities facilitate the introduction of carboxyl groups onto MEMS surfaces, enhancing biomolecule immobilization for biosensors and lab-on-chip platforms. The mild chemical reactivity of lactic acid makes it ideal for surface modification of delicate substrates without compromising structural integrity, advancing biomedical sensors and implantable MEMS technologies [2].

The etching of noble metals, particularly gold (Au), remains a significant challenge due to the toxicity and disposal difficulties of traditional cyanide-based etchants. To overcome these issues, iodine-iodide and thiosulfate-based etchants have been explored as safer alternatives. These systems dissolve Au by forming soluble Au iodide complexes, providing effective etching with reduced health and environmental risks [7]. Nonetheless, process isotropy, undercutting, and galvanic effects can still affect pattern fidelity and surface quality.

An innovative approach to noble metal etching is the use of organic aqua regia, as described by Lin et al., which combines organic solvents and acids to dissolve noble metals efficiently while reducing environmental hazards. This technique improves compatibility with flexible and bio-integrated MEMS devices and mitigates isotropic etching, though further optimization is necessary to improve etch rates, pattern uniformity, and process scalability [8].

Despite the environmental benefits of organic acid-based wet etching, several limitations hinder its widespread industrial application. Chief among these is the inherently slower etch rate compared to mineral acid or plasma-based etching, which constrains throughput and limits the practicality of organic acid etching for complex or high-aspect-ratio microstructures [6].

Temperature sensitivity is another critical factor affecting process performance. Although elevated temperatures can enhance etch rates by accelerating reaction kinetics and improving material solubility, excessive heat may compromise surface quality, increase lateral etching, and cause process instability. Studies on Ga-doped ZnO thin films and quartz substrates highlight the need for precise temperature control to ensure high-quality surfaces and minimal defects [2, 5].

To address these challenges, hybrid etching processes that combine organic acid chemistry with plasma, ultrasonic agitation, or catalyst-assisted methods have been developed. For instance, He et al. introduced a hybrid etching process that integrates wet etching with inductively coupled plasma reactive ion etching (ICP-RIE) for fabricating sub-micron type-II GaAsSb/InP double heterojunction bipolar transistors (DHBTs). This approach enhances device isolation, suppresses lateral undercutting, and improves uniformity, supporting the production of high-performance MEMS and high-frequency devices [9].

In metal etching, the combination of organic acids with diluted HF has been shown to reduce HF-induced surface roughness while preserving etch rates and dimensional control. Pernel et al. demonstrated that citric and oxalic acid formulations effectively mitigate the aggressive nature of HF, providing safer and more reliable copper etching processes for MEMS metallization and packaging applications [6].

Quartz and glass substrate etching using organic acid solutions has also shown promising results. Wan et al. reported that organic acid-based etching of quartz achieves high selectivity and generates less hazardous waste than conventional HF-based techniques. Such processes are critical for transparent MEMS, micro-optical devices, and bio-integrated microsystems, where precise patterning and minimal environmental impact are essential [2].

Ongoing research efforts focus on improving the performance of organic acid-based etching through process optimization, additive engineering, and waste management innovations. Catalyst-assisted etching, utilizing metal ions or nanoparticles, accelerates reaction rates and enhances etch uniformity without introducing hazardous reagents. The addition of surfactants and chelating agents into organic acid formulations further improves selectivity, surface quality, and isotropy [4, 5].

Industrial-scale implementation of organic acid etching requires effective waste management strategies. Although these processes inherently produce less hazardous waste than mineral acid-based systems, large-scale production necessitates closed-loop recycling and chemical recovery. Membrane separation, solvent extraction, and selective precipitation are promising methods for minimizing waste and enhancing the sustainability of organic acid-based etching [2].

Despite these advancements, material compatibility, etch rate enhancement, and process standardization remain ongoing challenges. The current limitations in processing certain metals, dielectrics, and multilayer structures restrict the broader applicability of organic acid etching in advanced MEMS and semiconductor manufacturing.

Looking forward, expanding the material scope of organic acid etching to include diverse functional materials used in MEMS, photonics, and flexible electronics remains a priority. The development of hybrid organic-inorganic etching chemistries, combined with plasma, laser-assisted, or electrochemical processes, is anticipated to unlock new capabilities for high-precision, environmentally responsible microfabrication [9].

In conclusion, organic acid-based wet etching represents a significant step toward greener, safer, and more precise MEMS manufacturing. Through continued innovation in process chemistry, hybrid etching integration, and waste reduction strategies, organic acid-based etching is poised to play a central role in the next generation of high-performance, eco-friendly MEMS and microfabrication technologies. A comparison of various organic acids for etching MEMS structures is listed in Table 5.1 [1–9].

TABLE 5.1

Technical Properties of Organic Acid-Based Etchants and Alternatives for MEMS Applications [1–9]

Etchant / System	Material Compatibility	Etch Selectivity	Surface Quality	Process Control	Advantages for MEMS	Limitations
Citric acid + H$_2$O$_2$	III-V semi-conductors (GaAs, AlGaAs, InP, InGaAs, InAlAs)	High selectivity (InGaAs/ InP: >400)	Good surface integrity with optimized conditions	Highly sensitive to acid concentration, oxidant ratio, temperature, agitation	Precise etching for optoelectronic & high-frequency MEMS, reduced damage	Requires tight process control, risk of surface roughness at high oxidant levels
Oxalic acid	Metals (copper), Ga-doped ZnO, transparent oxides	Moderate to High, depends on passivation effect	Excellent (Smooth, low-roughness surfaces)	Sensitive to chelating effect, needs precise control	High-dimensional precision, optical MEMS, display applications	Slower etch rates due to passivation, reduced throughput
Lactic acid	Delicate substrates (e.g., bioMEMS, microfluidics)	Not primarily used for etching, but for surface functional-ization	Preserves structural integrity, enables surface modi-fication	Mild chemical reactivity, biocompatible	Enhances biomolecule immobilization, ideal for biosensors, lab-on-chip	Limited as a primary etchant, mostly surface functional-ization role
Iodine-iodide, thiosulfate	Noble metals (gold)	Moderate, depends on formulation	Can be acceptable, though isotropy issues possible	Requires optimization to reduce undercutting & galvanic effects	Safer alternative to cyanide-based gold etchants	Process isotropy & pattern fidelity concerns remain
Organic aqua regia	Noble metals (gold, platinum)	High, with reduced isotropic etching	Improved over traditional mineral acid systems	Still under development, optimization needed	Efficient noble metal etching for flexible, bio-integrated MEMS	Requires further improvements for uniformity & scalability
Organic acid + diluted HF (hybrid)	Copper, other metals	Enhanced control vs. pure HF	Reduced HF-induced roughness	Combines benefits of organic acids & HF	Safer, reliable metal etching for MEMS metallization	Still involves HF, though in reduced concentrations
Organic acid for quartz/ glass	Quartz, glass substrates	High selectivity	Smooth, precise etching	Temperature-sensitive, requires careful control	Ideal for transparent MEMS, micro-optical devices	Lower etch rates than HF-based etching, limits scalability

5.2 SUPERCRITICAL CO$_2$-BASED ETCHING FOR MEMS

The continuous advancement of MEMS requires not only improved device performance but also greener and more sustainable microfabrication technologies. In recent years, supercritical carbon dioxide (scCO$_2$) has emerged as a highly promising alternative medium for etching, cleaning, and releasing delicate MEMS structures, owing to its unique physicochemical properties that enable residue-free, stiction-free, and environmentally responsible processing.

Supercritical CO$_2$ is a fluid state of carbon dioxide that occurs when both its temperature and pressure exceed critical values of 31.1°C and 73.8 bar, respectively. In this supercritical state, CO$_2$ exhibits excellent solvating power similar to that of a liquid, while retaining the diffusivity and viscosity of a gas. This unique combination enables scCO$_2$ to penetrate deep into high-aspect-ratio features, narrow gaps, and porous structures typical of MEMS devices, where conventional liquid-phase processing often fails or causes stiction-induced structural collapse [10, 11]. Furthermore, scCO$_2$ has near-zero surface tension, which eliminates the capillary forces that are primarily responsible for structure deformation during conventional wet processing (Figure 5.1).

The first and most established application of scCO$_2$ in MEMS processing is for stiction-free drying following sacrificial layer removal. Dyck et al. [12] demonstrated the successful use of scCO$_2$ solvent extraction for releasing surface-micromachined MEMS structures, effectively eliminating stiction and preserving structural integrity. Compared to traditional drying methods, which introduce liquid-vapor interfaces during evaporation, scCO$_2$ drying bypasses these interfaces altogether, providing a gentler and more reliable release process for fragile microstructures.

Building upon these early successes, scCO$_2$-based processes have evolved beyond drying to active roles in etching, sacrificial layer removal, and device cleaning. Bae et al. [13] reported an improved etching method for microelectronic devices utilizing scCO$_2$ as the reaction medium. Their approach leverages scCO$_2$'s superior transport properties to dissolve reactive species that enable precise, uniform, and residue-free etching of materials such as silicon dioxide (SiO$_2$) sacrificial layers. Their findings highlighted that scCO$_2$-based etching effectively mitigates challenges associated with conventional wet etching, such as uneven material removal, surface contamination, and process-induced damage.

One of the most significant advantages of scCO$_2$-based etching and cleaning processes is their excellent compatibility with delicate MEMS materials and structures.

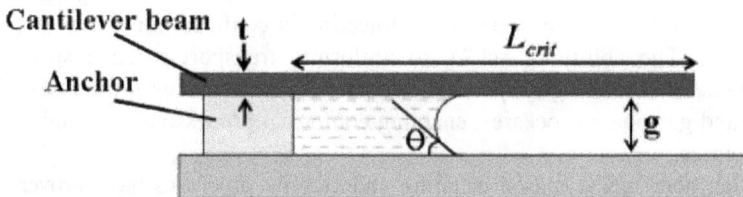

FIGURE 5.1 A cantilever beam of length l and thickness t anchored at an initial gap spacing g. The beam is attached to the substrate at distance L_{crit} from the anchor [10].

Aggressive wet etching chemicals such as HF or strong bases often degrade or damage low-k dielectrics, polymers, and porous materials commonly used in advanced MEMS devices. In contrast, $scCO_2$ leaves no liquid residue upon depressurization, minimizes chemical exposure, and does not compromise the structural integrity of fragile components [14, 15].

The compatibility of $scCO_2$ with porous and low-density materials has enabled its successful integration into MEMS processes where conventional wet or dry etching is problematic. Bouchaour et al. [16] investigated the role of $scCO_2$ in the drying of porous silicon, a material widely employed in MEMS sensors, optical devices, and microfluidics. Their work demonstrated that $scCO_2$ drying preserves the structural integrity and high surface area of porous silicon, avoiding capillary-induced damage that typically occurs with liquid-phase processing.

In parallel, Jung et al. [17] extended $scCO_2$-based etching to address the removal of tetraethyl orthosilicate (TEOS) oxide sacrificial layers for polycrystalline silicon (poly-Si) MEMS cantilevers. Their study confirmed that $scCO_2$-based etching achieves uniform oxide removal within high-aspect-ratio structures, while minimizing structural damage or undercutting that often occurs with conventional isotropic wet etching. This advancement represents a critical step toward reliable release processes for complex MEMS components.

Supercritical CO_2 is inherently chemically inert under most processing conditions, making it a safe and environmentally benign processing medium. However, its etching and cleaning capabilities are significantly enhanced through the introduction of co-solvents, reactive species, or surfactants tailored to specific material systems. For instance, Korzenski et al. [18] patented formulations that combine $scCO_2$ with chemical additives for the controlled removal of MEMS sacrificial layers. These approaches leverage $scCO_2$'s superior transport properties to deliver etchants uniformly into confined structures, facilitating efficient and selective material removal.

The environmental advantages of $scCO_2$ processing are considerable. Unlike traditional wet etching methods that generate large volumes of hazardous liquid waste and require complex waste treatment systems, $scCO_2$-based processes leave no liquid residue upon depressurization. CO_2 itself is non-toxic, non-flammable, and readily available as a byproduct of industrial operations, contributing to the sustainability and economic attractiveness of this technology [10].

Romang and Watkins [19] provided a comprehensive overview of supercritical fluids, emphasizing the versatility of $scCO_2$ in semiconductor and MEMS fabrication. They highlighted its potential not only for etching and cleaning but also for deposition, infiltration, and surface modification in confined micro- and nanoscale geometries. The ability of $scCO_2$ to uniformly transport reactive species into narrow channels and high-aspect-ratio features overcomes key limitations of both liquid- and gas-phase processes, enabling conformal processing in complex device architectures.

Furthermore, $scCO_2$-based cleaning and etching processes have proven effective for removing photoresist residues, polymer films, and other contaminants from MEMS devices without damaging sensitive surfaces. Saga and Hattori [15] demonstrated that $scCO_2$, when combined with suitable co-solvents, enables efficient photoresist stripping and cleaning of low-k dielectrics—an essential requirement for advanced semiconductor and MEMS applications.

These advancements illustrate that scCO$_2$-based etching is not merely a laboratory curiosity but a viable solution for scalable, high-yield MEMS manufacturing. Nevertheless, challenges remain, particularly in developing standardized processes, optimizing chemical formulations, and addressing equipment-related cost barriers.

In conclusion, supercritical CO$_2$-based etching and cleaning represent a powerful and environmentally sustainable alternative to conventional wet and dry etching techniques for MEMS fabrication.

5.3 IONIC LIQUID ETCHING FOR MEMS MATERIALS

The continuous evolution of MEMS demands precise, safe, and sustainable microfabrication processes. In this context, ionic liquid (IL)-based etching has emerged as a powerful and environmentally friendly alternative to conventional wet and dry etching techniques. ILs, defined as organic salts with melting points below 100°C, possess unique physicochemical properties such as negligible vapor pressure, excellent chemical tunability, high thermal stability, and broad electrochemical windows. These features make them particularly attractive for diverse MEMS fabrication processes, including etching, deposition, and surface modification.

A key motivation behind the development of IL-based etching is the need to replace hazardous etchants such as HF, which, while effective for etching silicon and glass, poses severe health, environmental, and handling risks. ILs, with their inherently low volatility and customizable chemical compositions, offer a safer alternative that aligns with global trends in green manufacturing [20–22].

Recent research has demonstrated the viability of ILs for silicon micromachining. Saverina et al. successfully fabricated porous silicon structures via electrochemical etching using imidazolium-based ILs such as [HMIM][BF4] and [HMIM][PF6] under mild, room-temperature conditions [23]. By fine-tuning critical parameters such as current density, etching time, IL composition, and water content, the researchers achieved uniform porous structures with pore sizes ranging from 30 to 80 nm. Notably, the optimal water content of approximately 1% significantly improved etch rates and surface quality while preserving the recyclability of the ILs. This work underscores the potential of IL-based etching for fabricating porous silicon, essential for optoelectronics, biosensors, and energy devices.

Beyond silicon, IL-based etching has been explored for metallic materials, notably aluminum, which plays a vital role in MEMS packaging, electrical interconnects, and actuation. Al Farisi et al. developed a patterned electroplating process for aluminum using AlCl3–[EMIm]Cl ILs, achieving high-quality, conformal metallic microstructures compatible with CMOS technologies [24]. Compared to conventional aqueous electroplating, this IL-based approach eliminates issues related to hydrogen evolution, improves process uniformity, and enables the fabrication of complex metallic features critical for advanced MEMS devices.

The unique electrochemical properties of ILs have also advanced the fabrication of micropropulsion components for space applications. Gassend et al. fabricated planar electrospray arrays using ILs as propellants, demonstrating stable ion emission at low voltages from MEMS-based emitters with integrated extractor components [25]. Such devices are crucial for nanosatellite propulsion, precise material deposition, and mass spectrometry. Similarly, Xue et al. developed tungsten emitters for ionic liquid electrospray thrusters (ILETs) using low-speed wire cutting combined with IL-based electrochemical etching [26]. The resulting emitters exhibited improved wettability, uniform IL film formation, and stable thrust performance in micropropulsion applications.

Glass micromachining, essential for MEMS packaging and microfluidic devices, presents additional opportunities for IL-based etching. Traditional HF-based glass etching is effective but hazardous. Choi et al. reviewed eco-friendly glass etching alternatives, emphasizing the potential of ILs and alkaline-based processes to safely replace HF for applications like through-glass vias (TGVs) and microfluidic channels [20]. These methods promote worker safety, reduce chemical waste, and align with sustainability goals in MEMS fabrication.

Fundamentally, ILs enable precise control over etching processes due to their tunable chemistry and compatibility with electrochemical techniques. Gao et al. highlighted the ability to design superwetting surfaces using ILs, which directly benefits IL-based etching by enhancing wettability, reducing surface defects, and improving etch uniformity [21]. Such tailored IL formulations enable fine control over etching rates, selectivity, and surface morphology, facilitating the fabrication of complex MEMS architectures.

In addition to silicon and glass, ILs have shown promising potential for etching ceramic materials, which are increasingly used in MEMS for their excellent mechanical and thermal properties. Hossain et al. demonstrated microfabrication of ceramic-based MEMS using IL-assisted micromachining, achieving high-aspect-ratio structures with minimal surface damage [27]. Their work highlights ILs' ability to process hard and brittle materials with superior control compared to traditional etchants.

Recent advancements also include the use of ILs for the etching and functionalization of nanostructured surfaces. Gao et al. reported IL-based strategies for modifying micro- and nanostructures with superhydrophobic or superhydrophilic properties, which can directly impact MEMS performance in fields such as microfluidics and sensing [21]. Such functionalization expands IL-based etching beyond structural fabrication to the engineering of device surfaces.

Despite these advantages, several challenges remain in scaling IL-based etching for commercial MEMS production. Process optimization is needed to improve etch rates, selectivity, and reproducibility across diverse materials and device geometries. Impurities, water content, and IL degradation during extended use can compromise process stability and must be carefully managed. Recycling and purification protocols are essential to ensure IL sustainability and economic viability [22].

Emerging research focuses on task-specific ILs designed for targeted MEMS etching applications. Functionalized ILs incorporating reactive species, chelating groups, or surfactants can enhance etch performance, selectivity, and recyclability. For instance, IL mixtures with controlled acidity or basicity have shown promise for silicon, metal, ceramic, and glass etching, providing tunable reaction kinetics and improved material compatibility [23, 24, 27].

The integration of IL-based etching with hybrid microfabrication processes offers additional pathways to overcome current limitations. Plasma-assisted IL etching, for example, combines the precision of plasma processes with the chemical tunability of ILs, enabling anisotropic etching and high aspect ratio structure formation. Electrochemical patterning using ILs also shows potential for selective material removal and localized microstructuring in MEMS devices [25].

Furthermore, IL-based etching aligns with broader trends toward additive and subtractive manufacturing of multifunctional MEMS. The ability to perform both material removal and deposition in IL media simplifies process integration and enables novel device architectures. Applications extend to bioMEMS, flexible electronics, and microfluidics, where IL compatibility with soft materials, biological interfaces, and complex 3D geometries provides significant advantages [26]. Ongoing research continues to explore IL interactions with emerging MEMS materials such as graphene, 2D materials, and functional ceramics. IL-based processes for etching, doping, and functionalizing these materials hold promise for next-generation MEMS sensors, energy harvesters, and optoelectronic devices [27].

In conclusion, IL-based etching represents a promising, versatile, and sustainable approach to MEMS microfabrication. Through careful IL selection, process optimization, and integration with advanced microfabrication techniques, IL etching is poised to significantly contribute to safer, greener, and more precise MEMS manufacturing. Continued research and development will unlock the full potential of ILs in addressing the challenges of next-generation MEMS device fabrication.

5.4 HYBRID WET ETCHING AND ELECTROCHEMICAL ETCHING FOR MEMS

The continued evolution of MEMS fabrication has intensified the demand for microfabrication techniques that not only deliver high precision and versatility but also meet sustainability and environmental safety standards. Conventional wet etching, despite its simplicity and low cost, often involves the use of hazardous chemicals like HF, HNO_3, and strong alkaline solutions. These substances pose significant health, safety, and environmental risks, particularly as MEMS devices become more intricate and production scales increase. In response, hybrid wet etching and electrochemical etching have emerged as promising alternatives, combining the strengths of multiple etching approaches to achieve precise, controllable, and eco-friendly MEMS fabrication.

One of the most widely adopted hybrid etching techniques is metal-assisted chemical etching (MACE), which leverages noble metal catalysts in conjunction with wet etching to produce high-aspect-ratio microstructures in silicon and other materials. In the conventional MACE process, a thin metal layer—commonly Au, silver (Ag), or platinum (Pt)—is deposited onto a silicon substrate patterned via photolithography. When immersed in an etching solution containing an oxidizing agent such as H_2O_2 and HF, the metal catalyzes localized oxidation of silicon beneath the metal layer, which is subsequently dissolved by HF, resulting in anisotropic etching and near-vertical sidewalls [9, 28].

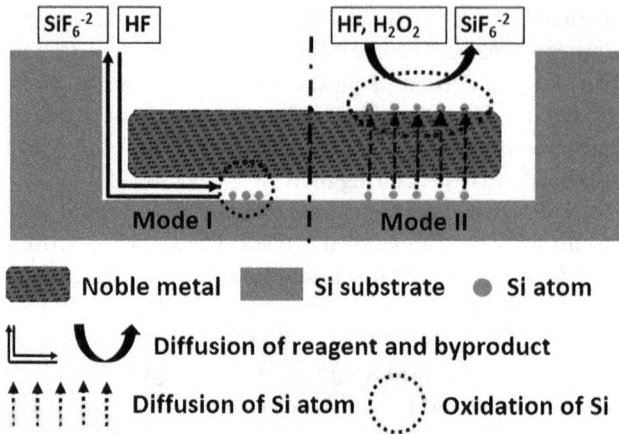

FIGURE 5.2 Schematic diagram of metal-assisted chemical etching (MACE). H_2O_2 and HF as reactants and the metal layer as a catalyst. The reduction of H_2O_2 produces holes, which then oxidize the Si layer. (Source: Ref. [29].)

Recent advancements have significantly enhanced the applicability and efficiency of MACE for MEMS fabrication. Nur'aini and Oh demonstrated a deep etching method based on MACE that overcomes the traditional limitation of mass transport, which previously confined MACE primarily to nanostructures (Figure 5.2) [29]. By utilizing an ultrathin metal layer below 10 nm, the out-of-plane mass transfer at the metal/silicon interface was promoted, enabling the etching of micrometer-scale structures with vertical profiles exceeding 85° and etch depths greater than 200 μm at rates over 0.4 μm per minute. The use of ethanol as a solvent, rather than water, further suppressed porous defect formation, resulting in high-quality, defect-free microstructures suitable for MEMS and biosensor applications.

The stability of the metal catalyst plays a crucial role in determining etch quality and depth. Single Au thin films have demonstrated superior stability and etch uniformity compared to Ag or Ag/Au bilayer catalysts, while also eliminating issues of catalyst corrosion and porous defect formation. Optimization of solvent composition, oxidant concentration, and metal layer thickness is a key factor in achieving the desired etch profile and minimizing defects [29].

Electrochemical etching offers another effective strategy for precision MEMS microstructuring, particularly for creating porous silicon layers, high-aspect-ratio trenches, and freestanding structures. By applying an electrical bias to the substrate immersed in an electrolyte, localized electrochemical reactions can be controlled to selectively remove material. Yang and colleagues demonstrated that electrochemical etching provides tunable porosity, precise depth control, and reduced HF consumption, making it ideal for producing porous silicon layers essential for gas sensors, thermal isolation structures, and other MEMS components [30].

The combination of electrochemical etching with MACE has enabled the fabrication of hierarchical silicon structures with enhanced surface area, mechanical stability, and functional versatility, expanding their application to MEMS gas sensors, biosensors, supercapacitors, and energy harvesters [9, 29].

For III-V compound semiconductors such as GaAs and InP, hybrid etching processes that integrate wet chemical etching with ICP-RIE have demonstrated significant benefits. He and colleagues developed a hybrid wet and ICP-RIE process for fabricating sub-micron Type-II GaAsSb/InP DHBTs, achieving superior device isolation, minimized lateral undercutting, and improved process uniformity suitable for high-performance MEMS and millimeter-wave applications [9].

Similarly, the integration of citric acid-based electrolytes into electrochemical etching processes offers a more environmentally friendly alternative to conventional HF-based etching of III-V semiconductors. DeSalvo and co-workers reported that citric acid–H_2O_2 mixtures can selectively etch these materials with excellent surface quality, highlighting their potential for optoelectronic and high-frequency MEMS applications [3].

Glass and ceramic materials, integral to microfluidics, bioMEMS, and optical MEMS, have also benefited from hybrid etching developments. Choi and co-workers emphasized the potential of metal-assisted and hybrid etching processes to replace hazardous HF-based techniques for glass micromachining while maintaining structural precision and process efficiency [1]. These approaches, which incorporate MACE, electrochemical etching, and optimized wet chemistries, have been shown to improve etch uniformity, surface smoothness, and feature definition, advancing the fabrication of transparent MEMS, optical devices, and bio-integrated microsystems.

Hybrid wet and electrochemical etching techniques offer significant environmental advantages by reducing hazardous chemical consumption, improving etch efficiency, and enabling waste minimization. For instance, optimized MACE processes can reduce HF concentrations by up to 50%, while the use of biodegradable organic acids in electrochemical etching further enhances sustainability [31, 32].

However, challenges related to scalability, reproducibility, and process uniformity persist, particularly for large-area or high-volume MEMS production. Non-uniform catalyst deposition, electrolyte flow variations, and electric field distribution inconsistencies can lead to significant process variability, limiting industrial adoption. Nur'aini and Oh addressed these challenges by demonstrating that optimized metal layer configurations, solvent selection, and controlled oxidant concentrations enable the fabrication of high-aspect-ratio, defect-free microstructures suitable for MEMS at larger scales [29].

Continued research efforts are needed to refine hybrid wet and electrochemical etching processes for MEMS fabrication. The development of task-specific electrolytes and etching chemistries tailored to diverse MEMS materials and device architectures remains critical. The integration of machine learning and artificial intelligence for real-time process monitoring and predictive control offers a promising route to enhance etch precision and yield. Exploration of novel solvent systems, catalyst materials, and additive engineering could further suppress defects and improve etch profiles. Furthermore, advancements in closed-loop chemical recycling and waste management systems will be essential to minimize the environmental impact of these processes.

In conclusion, hybrid wet etching and electrochemical etching represent essential tools for advancing high-performance, sustainable MEMS fabrication. By integrating complementary etching mechanisms and leveraging recent innovations in material science and process engineering, these approaches offer unparalleled control over feature definition, material selectivity, and environmental safety. As these technologies mature, they are poised to play a central role in enabling the next generation of eco-friendly, reliable, and versatile MEMS devices.

REFERENCES

1. Choi, K., Kim, S. W., Lee, J. H., Chu, B., & Jeong, D. Y. (2024). Eco-friendly glass wet etching for MEMS application: A review. *Journal of the American Ceramic Society*, 107, 6497–6515.
2. Wan, Y., Luan, X., Zhou, L., & Wu, F. (2022). Wet etching of quartz using a solution based on organic solvents and anhydrous hydrofluoric acid. *Materials*, 15(18), 6475.
3. DeSalvo, G. C., Tseng, W. F., & Comas, J. (1992). Etch rates and selectivities of citric acid/hydrogen peroxide on GaAs, $Al_{0.3}Ga_{0.7}As$, $In_{0.2}Ga_{0.8}As$, $In_{0.53}Ga_{0.47}As$, $In_{0.52}Al_{0.48}As$, and InP. *Journal of the Electrochemical Society*, 139(3), 831–837.
4. Xia, Y., Zhao, X. M., Kim, E., & Whitesides, G. M. (1995). A selective etching solution for use with patterned self-assembled monolayers of alkanethiolates on gold. *Chemistry of Materials*, 7(12), 2332–2337.
5. Lee, D. K., Lee, S. J., Bang, J., & Yang, H. (2008). Wet etching behaviors of transparent conducting Ga-doped zinc oxide thin film by organic acid solutions. *IMID/IDMC/Asia Display 2008 Digest*, 8, 831–833.
6. Pernel, C., Farkas, J., & Louis, D. (2006). Copper in organic acid based cleaning solutions. *Journal of Vacuum Science & Technology B*, 24(5), 2467–2471.
7. Green, T. A. (2014). Gold etching for microfabrication. *Gold Bulletin*, 47(4), 205–216.
8. Lin, W., Zhang, R. W., Jang, S. S., Wong, C. P., & Hong, J. I. (2010). Organic aqua regia—powerful liquids for dissolving noble metals. *Angewandte Chemie International Edition*, 49(46), 7929–7932.
9. He, Y., Peng, Y. T., Yu, X., & Feng, M. (2022). Hybrid etching process in sub-micron type-II GaAsSb/InP DHBT for 5G and millimeter-wave power amplification. In: *Proceedings of CS MANTECH Conference*, pp. 1–4.
10. Hwang, H. S., Bae, J. H., Jung, J. M., & Lim, K. T. (2010). The sacrificial oxide etching of poly-Si cantilevers having high aspect ratios using supercritical CO_2. *Microelectronic Engineering*, 87, 1696–1700.
11. Weibel, G. L., & Ober, C. K. (2003). An overview of supercritical CO_2 applications in microelectronics processing. *Microelectronic Engineering*, 65, 145–152.
12. Dyck, C. W., Smith, J. H., Miller, S. L., Russick, E. M., & Adkins, C. L. J. (1996). Supercritical carbon dioxide solvent extraction from surface-micromachined micromechanical structures. *SPIE Micromachining and Microfabrication Fabrication Process Technology II*, 2879, 225–235.
13. Bae, J. H., Alam, M. Z., Jung, J. M., Gal, Y. S., Lee, H., Kim, H. G., & Lim, K. T. (2009). Improved etching method for microelectronic devices with supercritical carbon dioxide. *Microelectronic Engineering*, 86, 128–131.
14. Iyer, S. S. et al., 2004. Etching and cleaning of semiconductor devices using supercritical fluids. US Patent US20040134885A1.
15. Saga, K., & Hattori, T. (2007). Wafer cleaning using supercritical CO_2 in semiconductor and nanoelectronic device fabrication. *Solid State Phenomena*, 134, 97–102.
16. Bouchaour, T., Diaf, N., Ould-Abbas, A., Benosman, M., Merad, L., & Chabane-Sari, N. E. (2003). The role of supercritical CO_2 in the drying of porous silicon. *Revue des Energies Renouvelables*, ICPWE, 99–102.
17. Jung, J. M., Yoon, E., Lim, E., Choi, B. C., Kim, S. Y., & Lim, K. T. (2011). The dry etching of TEOS oxide for poly-Si cantilevers in supercritical CO_2. *Microelectronic Engineering*, 88, 3448–3451.

18. Korzenski, M. et al. (2003). Removal of MEMS sacrificial layers using supercritical fluids/chemical formulation. US Patent US20050118832A1.
19. Romang, A. H., & Watkins, J. J. (2010). Supercritical fluids for the fabrication of semiconductor devices: Emerging or missed opportunities? *Chemical Reviews*, 110, 459–478.
20. Xu, T., Tao, Z., & Lozano, P. C. (2018). Etching of glass, silicon, and silicon dioxide using negative ionic liquid ion sources. *Journal of Vacuum Science & Technology B*, 36, 052601.
21. Gao, N., Zhou, K., Feng, K., Zhang, W., Cui, J., Wang, P., Tian, L., Jenkinson-Finch, M., & Li, G. (2021). Facile fabrication of self-reporting micellar and vesicular structures based on an etching-ion exchange strategy of photonic composite spheres of poly(ionic liquid). *Nanoscale*, 13, 1927–1936.
22. Storey, A. C. G., Sabouri, A., Khanna, R., & Ahmed, U. (2024). Filtering the beam from an ionic liquid ion source. *Journal of Vacuum Science & Technology B*, 42, 064201.
23. Saverina, E. A., Zinchenko, D. Y., & Farafonova, S. D., et al. (2020). Porous silicon preparation by electrochemical etching in ionic liquids. *ACS Sustainable Chemistry & Engineering*, 8(27), 10023–10033.
24. Al Farisi, M. S., Hertel, S., Wiemer, M., & Otto, T. (2018). Aluminum patterned electroplating from AlCl$_3$–[EMIm]Cl ionic liquid towards microsystems application. *Micromachines*, 9(11), 589.
25. Gassend, B., Velásquez-García, L. F., Akinwande, A. I., & Martínez-Sánchez, M. (2009). A microfabricated planar electrospray array ionic liquid ion source with integrated extractor. *Journal of Microelectromechanical Systems*, 18(3), 679–693.
26. Xue, S., Duan, L., & Kang, Q. (2021). Fabrication of externally wetted emitter for ionic liquid electrospray thruster by low-speed wire cutting combined with electrochemical etching. *AIP Advances*, 11(11), 115023.
27. Hossain, K. R., Jiang, P., Yao, X., Yang, X., Hu, D., & Wang, X. (2023). Ionic liquids for 3D printing: Fabrication, properties, applications. *Journal of Ionic Liquids*, 3, 100066.
28. Li, X. (2012). Metal-assisted chemical etching for high-aspect-ratio nanostructures: A review of characteristics and applications in photovoltaics. *Current Opinion in Solid State and Materials Science*, 16(2), 71–81.
29. Nur'aini, A., & Oh, I. (2022). Deep etching of silicon based on metal-assisted chemical etching. *ACS Omega*, 7(19), 16665–16669.
30. Yang, C. R., Fu, P. C., Cheng, C., & Huang, M. J. (2021). An integrated microelectromechanical system-based silicon wet etching process and nanocarbon materials used for improving micro direct methanol fuel cells performance. *Materials Today Energy*, 20, 100696.
31. Jung, K., and Lee, J. (2024), A review of the mechanism and optimization of metal-assisted chemical etching and applications in semiconductors. *Micro and Nano Systems Letters*, 12, 27
32. Akan, R and Vogt, U. (2021). Optimization of Metal-Assisted Chemical Etching for Deep Silicon Nanostructures. *Nanomaterials*, 11(11), 2806.

6 Electrochemical Etching for Green MEMS Fabrication

6.1 FUNDAMENTALS OF ELECTROCHEMICAL ETCHING IN MEMS

Electrochemical etching (ECE) has emerged as a key technique for precise and selective material removal in microelectromechanical systems (MEMS) fabrication. As the demand for sustainable and environmentally conscious manufacturing processes grows, ECE offers significant advantages over conventional dry and wet etching techniques, including low energy consumption, reduced use of hazardous chemicals, and excellent material selectivity. In the context of green MEMS fabrication, ECE provides a promising pathway to realize complex micro- and nanostructures in a manner aligned with environmental sustainability principles.

The fundamental principle of ECE is based on the controlled dissolution of semiconductor materials through electrochemical reactions at the interface between the material surface and a liquid electrolyte. This process typically requires the application of an external electrical bias, which drives the oxidation of the semiconductor at the anode, followed by the removal of the oxidized species into the electrolyte. For silicon, which remains the most widely used material in MEMS fabrication, the anodic dissolution process in hydrofluoric acid (HF)–based electrolytes follows the well-known reaction:

$$Si + 6HF \rightarrow H_2SiF_6 + H_2 + 2H^+ + 2e^-$$

Through careful control of the applied voltage, current density, electrolyte composition, and process conditions, the etch rate, morphology, and selectivity of the process can be precisely tuned to meet the requirements of various MEMS applications. This approach has been extensively applied in porous silicon fabrication, microstructure release, and selective layer removal in silicon-based microsystems [1].

However, despite the technical advantages of HF-based ECE processes, the use of HF poses significant safety and environmental hazards. HF is highly toxic, corrosive, and environmentally persistent, leading to considerable efforts to develop alternative green chemistries for ECE. Recent advances have demonstrated that alkaline electrolytes, such as potassium hydroxide (KOH), can be effectively used for the ECE of certain semiconductor materials, reducing the ecological footprint of the process [2]. Additionally, localized etching strategies that minimize electrolyte consumption have been explored as an effective route to green microfabrication [3].

ECE has proven particularly advantageous for structuring wide bandgap semiconductor materials such as silicon carbide (SiC), which offer superior mechanical,

DOI: 10.1201/9781003674795-6

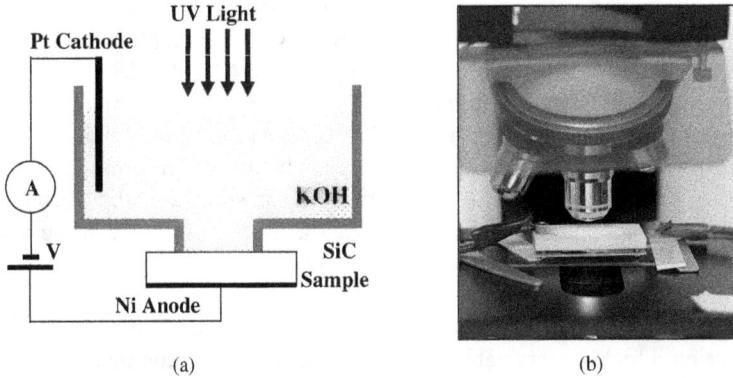

FIGURE 6.1   (a) Schematic and (b) micrograph of the PEC etch setup [4].

thermal, and chemical stability compared to conventional silicon but are notoriously difficult to etch using traditional methods. In a landmark study, Zhao et al. demonstrated the use of dopant-selective photoelectrochemical (PEC) etching to fabricate single-crystal SiC MEMS structures. The schematic and photo of the PEC etch setup are shown in Figure 6.1 [4]. By exploiting the difference in flat-band potentials between n-type and p-type SiC in a KOH solution under illumination, selective undercutting of the p-SiC layer was achieved, resulting in freestanding microstructures with exceptional mechanical robustness. This approach enabled the realization of MEMS components capable of operating in harsh environments, including high-temperature and corrosive conditions.

Building on these principles, Hochreiter et al. developed an innovative ECE strategy for fabricating monolithic three-dimensional structures from 4H-SiC wafers [5]. In this work, etchable regions were defined by localized implantation of p-type dopants, which introduced significant contrasts in electrochemical reactivity between doped and undoped areas of the wafer. This doping-controlled selectivity allowed for the precise etching of complex structures, including cantilevers, disk-shaped resonators, and membranes, directly from single-crystal SiC substrates. The resulting structures exhibited high surface quality and remarkable mechanical stability, maintaining structural integrity at temperatures exceeding 1500°C. This approach eliminates the need for complex bonding or layer transfer processes and significantly expands the scope of MEMS devices suitable for operation in extreme environments.

ECE has also been successfully extended to III-nitride semiconductors, which are widely used in optoelectronic and high-power electronic devices. Han et al. reported the development of conductivity-selective lateral ECE of gallium nitride (GaN) and related materials for microfabrication applications [6]. This technique enables the layer-selective removal of sacrificial layers in multilayer GaN heterostructures, facilitating the formation of air-gap distributed Bragg reflectors (DBRs) essential for vertical-cavity surface-emitting lasers (VCSELs). The ability to selectively etch specific layers within complex nitride semiconductor structures has significant implications for both MEMS and photonic device fabrication, offering a scalable and environmentally conscious alternative to conventional dry etching techniques.

In recent years, significant progress has been made in developing green ECE methods that minimize chemical waste and energy consumption. A particularly innovative example is the triboelectric nanogenerator (TENG)-driven ECE system introduced by Wang et al. [3]. In this self-powered system, mechanical energy from the environment is harvested by a TENG to generate electrical output sufficient to drive localized ECE and deposition processes. The etching reaction is confined to a single droplet of electrolyte, which is precisely positioned using a needle electrode platform. By adjusting parameters such as droplet size, electrode distance, and TENG output, selective etching of aluminum and deposition of nanostructured materials such as silver and copper oxide were achieved. This approach eliminates the need for external power supplies and significantly reduces electrolyte consumption, offering a highly sustainable and cost-effective fabrication method for micro- and nanoscale devices.

The precision of ECE has also been significantly enhanced through the integration of passivation-assisted etch-stop strategies. Hong et al. developed an electrochemical passivation approach for controlling the fabrication of silicon nanopores with sub-10 nm feature sizes [7]. In this work, rapid and reversible electrochemical oxide growth was employed as a dynamic etch-stop mechanism within a three-step wet etching process. By precisely controlling the passivation potential, the researchers achieved highly accurate termination of the etching process within seconds, enabling the production of silicon nanopores with exceptional size control and uniformity. This technique is fully compatible with MEMS and nanofluidic applications and demonstrates the potential of ECE for high-density, high-precision nanostructuring in a scalable and environmentally responsible manner.

The application of ECE is not limited to semiconductors alone. The combination of ECE with green photoresist technologies, such as ADEX dry film photoresists, further enhances the environmental credentials of MEMS microfabrication. Roos et al. demonstrated the use of ADEX dry films for 3D microfabrication of bio-MEMS devices, eliminating the need for solvent-based lithography processes while providing excellent patterning capabilities and mechanical stability [8]. The integration of such green patterning materials with ECE offers a comprehensive approach to sustainable MEMS manufacturing, reducing chemical waste and energy consumption across multiple process steps.

Efforts to promote green MEMS fabrication extend beyond semiconductor and polymer structuring to include the etching of glass substrates for microfluidic and MEMS applications. Glass offers excellent chemical stability, optical transparency, and compatibility with biological environments, making it a popular choice for microfluidic devices and MEMS components. However, conventional glass wet etching often relies on HF, raising environmental and safety concerns. Recent studies by Choi et al. have reviewed alternative eco-friendly glass etching processes, including alkaline etching and innovative patterning methods, to address these challenges [2]. Such advances complement ECE by enabling sustainable fabrication of glass-based MEMS and microfluidic devices.

Despite its numerous advantages, ECE presents several challenges that must be addressed to fully realize its potential for green MEMS fabrication. The process often requires specific doping profiles or material heterostructures to achieve high selectivity, which may impose design constraints. Additionally, etch rates are typically lower than those achieved by aggressive plasma etching techniques, which can limit throughput in high-volume manufacturing. Maintaining uniformity across large substrate areas and ensuring reproducibility in complex structures remain active areas of research.

Nevertheless, ECE continues to evolve as a key enabling technology for sustainable MEMS fabrication. Through advances in doping-controlled selectivity, passivation-assisted precision, and integration with self-powered systems and green lithography techniques, ECE offers a unique combination of precision, material versatility, and environmental compatibility. Its ongoing development aligns with the global imperative for sustainable manufacturing, making it an indispensable component of the next generation of MEMS and microfabrication technologies.

6.2   ELECTROCHEMICAL ETCHING OF SILICON FOR MEMS

ECE has established itself as one of the most versatile and precise techniques for structuring silicon in MEMS fabrication. Compared to conventional dry etching or aggressive wet chemical methods, ECE offers a distinctive set of advantages, including excellent control over etching profiles, compatibility with complex three-dimensional microstructures, and process conditions that align well with the principles of environmentally conscious manufacturing. Through continuous technological advancement, ECE has progressed from its origins in porous silicon formation to a highly refined micromachining process capable of producing high-aspect-ratio MEMS structures with minimal ecological impact.

The ability to exploit electrochemical reactions at the silicon–electrolyte interface under controlled electrical bias forms the foundation of this technique. By carefully adjusting substrate doping, electrolyte composition, applied potential, and process duration, it is possible to achieve selective material removal with remarkable precision. In particular, n-type silicon under rear-side illumination has been extensively employed to facilitate uniform carrier generation, enabling deep vertical etching while preserving lateral dimensions.

One of the major breakthroughs in the application of ECE for MEMS fabrication was the development of electrochemical micromachining (ECM). Bassu et al. demonstrated how ECM allows for the fabrication of intricate freestanding MEMS components, such as inertial masses, comb-finger electrodes, and serpentine suspension springs, through a purely liquid-phase process without reliance on plasma-based dry etching techniques [9]. Their approach begins with pre-patterning the silicon substrate using oxide masking combined with KOH etching to define seed structures. Following this, the substrate undergoes a two-stage ECE process, consisting of an initial anisotropic etching phase to generate vertical structures and a subsequent isotropic phase to selectively undercut and release the microstructures from the substrate. By fine-tuning parameters such as current density, electrolyte concentration, and etching time, the structural anisotropy and etch rates can be precisely controlled, enabling the formation of high-aspect-ratio microstructures with excellent uniformity and reproducibility.

Compared to conventional deep reactive ion etching (DRIE), ECM offers significant advantages in terms of reduced equipment requirements, lower energy consumption, and the elimination of plasma-induced surface damage. Furthermore, the use of aqueous-based electrolytes under mild temperature conditions significantly reduces the environmental burden associated with the process. Through the integration of surfactants such as sodium lauryl sulfate, etch uniformity and surface quality have been further improved, contributing to the overall performance and reliability of MEMS devices produced by ECM [9].

Beyond bulk silicon micromachining, ECE has also proven highly effective for the fabrication of MEMS structures on silicon-on-insulator (SOI) substrates. SOI technology has become indispensable in MEMS fabrication due to its superior mechanical isolation, electrical insulation, and structural versatility. However, the release and operation of SOI-based MEMS devices often suffer from stiction, an unwanted adhesion phenomenon between movable and stationary components, which severely limits device functionality and reliability.

To address this persistent issue, Hao and co-workers developed a selective ECE technique that enables independent control over the etching of different layers within an SOI substrate [10]. Their method employs a dual-bias approach, where the device layer is held at a potential that promotes passivation, while the handle layer is selectively etched through controlled electrochemical dissolution. This process increases the gap between movable and stationary structures and introduces a degree of controlled surface roughness, significantly mitigating stiction during both fabrication and device operation. The practical benefits of this technique were demonstrated through the fabrication of cantilever-type MEMS switches. The process allowed for the fabrication of devices with detachment lengths exceeding 500 micrometers, while the tunable surface roughness on the handle layer effectively reduced the likelihood of stiction-induced failures. Importantly, this approach does not require doping variations within the silicon layers, simplifying substrate preparation and broadening the applicability of ECE for complex SOI-MEMS designs.

A critical factor in the successful application of ECE is the ability to predict and control etch profile evolution during processing. Given the complex interplay of electrical field distribution, mass transport, and reaction kinetics, numerical modeling has become an indispensable tool for understanding and optimizing the ECE process. Ivanov and Mescheder developed a dynamic simulation framework that incorporates both electrical and diffusion-driven mechanisms governing silicon anodization [11]. Their models confirmed that the initial stages of ECE are characterized by the well-known "edge effect," where electric field concentration near mask openings promotes the formation of convex etch profiles. As etching progresses, diffusion-limited transport of reactive species, particularly fluoride ions, becomes the dominant factor, leading to a transition from convex to concave or isotropic etch profiles. The models further demonstrated how electrolyte conductivity, applied voltage, and mask geometry directly influence the shape and uniformity of the resulting microstructures. High electrolyte conductivity was shown to enhance the edge effect and promote deeper vertical etching, while low conductivity conditions favored more uniform, isotropic etching profiles. These insights enable engineers to tailor process parameters to specific design requirements, minimizing defects and improving reproducibility.

In addition to enhancing process precision, simulation-driven optimization contributes significantly to reducing the environmental impact of ECE. By minimizing experimental trial-and-error, modeling decreases material waste, shortens development time, and reduces chemical consumption, aligning the process with green manufacturing objectives. Despite its numerous advantages, ECE has traditionally relied on HF as the primary electrolyte for silicon etching. While effective, HF poses significant safety risks, is highly toxic, and presents considerable challenges for waste management. Recognizing these environmental concerns, research efforts have increasingly focused on developing alternative electrolytes and process modifications to reduce or eliminate HF usage. Organic solvents, ionic liquids, and alkaline-based electrolytes have all been explored as potential substitutes, offering the possibility of achieving selective silicon dissolution under electrochemical conditions with lower toxicity and improved environmental profiles. Although these approaches have yet to fully replicate the selectivity and efficiency of HF-based processes, they represent a promising step toward greener silicon micromachining techniques [3, 9].

In parallel with electrolyte innovations, the development of localized and self-powered ECE systems has emerged as a transformative advancement in green MEMS fabrication. Wang et al. introduced a droplet-confined ECE platform powered by TENGs, which convert mechanical energy from the environment into electrical output sufficient to drive localized silicon etching and electrochemical deposition (ECD) [3]. By confining the reaction environment to microliter-scale droplets, the system dramatically reduces electrolyte consumption and eliminates the need for external power supplies. This approach not only simplifies equipment requirements but also supports portable, decentralized MEMS fabrication with minimal environmental footprint. Such localized ECE platforms open new avenues for resource-limited settings, field-deployable microsystem production, and on-demand prototyping. By enabling environmentally friendly, scalable, and low-energy fabrication, droplet-confined ECE systems complement broader trends toward distributed and sustainable microfabrication.

Another promising strategy for enhancing the environmental compatibility of silicon micromachining lies in the integration of ECE with green lithographic techniques. Conventional photolithography relies heavily on organic solvents, photoresist developers, and hazardous chemicals, which contribute to process waste and operator exposure risks. To overcome these limitations, solvent-free patterning techniques have been developed, with ADEX dry film photoresists offering one of the most effective solutions. Roos and co-workers demonstrated that ADEX resists enable high-resolution, scalable patterning suitable for both MEMS and bio-MEMS applications. When combined with ECE, these materials facilitate the production of complex microstructures while minimizing hazardous chemical use [8]. The solvent-free nature of ADEX resists, combined with the low-temperature operation of ECE, provides a highly compatible and environmentally favorable process chain. This is particularly advantageous for biomedical MEMS, where biocompatibility, process cleanliness, and reduced environmental impact are critical requirements. By integrating green lithography with ECE, high-performance microsystems can be fabricated with reduced environmental burden, improved process safety, and enhanced suitability for sensitive biological environments.

The future of ECE in MEMS fabrication lies in its continued evolution toward even greater precision, selectivity, and environmental sustainability. The development of

HF-free electrolyte formulations, building upon the promising results obtained with organic solvents and ionic liquids, remains a critical research direction. Further advancements in droplet-confined and self-powered ECE systems will likely expand the accessibility of microsystem production while minimizing resource consumption. Process modeling, incorporating machine learning and real-time process monitoring, is expected to play an increasingly important role in optimizing ECE processes for both performance and sustainability. The integration of ECE with additive manufacturing techniques and hybrid fabrication approaches also holds potential for expanding design flexibility, reducing material waste, and enabling new classes of MEMS devices.

ECE of silicon has proven itself as a cornerstone of high-precision, environmentally conscious MEMS fabrication. Its versatility, coupled with ongoing innovations in green chemistry, localized processing, and process integration, ensures that ECE will continue to play a central role in the realization of next-generation microsystems that meet the highest standards of functionality, reliability, and environmental responsibility.

6.3 ELECTROCHEMICAL ETCHING OF METALS AND CONDUCTIVE OXIDES

ECE has become an essential technique for the precise structuring, surface modification, and functionalization of metals and conductive oxides. Compared to conventional wet chemical or plasma-based etching methods, ECE offers significant advantages in terms of controllability, material selectivity, and environmental compatibility. As MEMS technology increasingly adopts diverse material platforms beyond semiconductors, the role of ECE in structuring metals and conductive oxides has expanded considerably. At the same time, its inherent alignment with green fabrication principles makes it a promising pathway toward more sustainable microsystem production.

The underlying principle of ECE for metals and conductive oxides relies on anodic dissolution, where metal atoms at the electrode–electrolyte interface undergo oxidation and are subsequently removed into the electrolyte solution. The rate and morphology of the etching process depend strongly on factors such as the applied potential, current density, electrolyte composition, and temperature. Through careful optimization of these parameters, ECE can be tuned to produce either highly uniform, isotropic etching for surface finishing or more localized, anisotropic etching for structural definition.

A particularly important development has been the application of ECE for the nanometric finishing of metal surfaces. Yi et al. demonstrated that by introducing sulfuric acid into the electrolyte, the mass transfer resistance during ECE of metals such as titanium, tungsten, and nickel–titanium alloys can be significantly increased. This modification promotes isotropic etching behavior, leading to uniform material removal across the surface. Their method, referred to as isotropic etching polishing (IEP), enabled the achievement of sub-10-nanometer surface roughness, a level of precision well-suited to high-performance metal components in MEMS, sensors, and biomedical devices. Compared to mechanical polishing or aggressive chemical etching, IEP provides a controllable, scalable, and environmentally favorable approach for producing ultrasmooth metallic surfaces without introducing micro-structural damage or chemical contamination [12].

In addition to material removal, ECE has proven highly effective for engineering functional surface structures that enhance properties such as corrosion resistance, hydrophobicity, and electrochemical activity. Wu and co-workers developed a green, wire-based ECE method to fabricate superhydrophobic nickel surfaces. Figure 6.2 shows the wire ECE process setup and related simulation results. By applying a localized cathodic potential through a wire electrode sweeping across the nickel sub-strate immersed in a sodium chloride solution, they produced micro- and nanoscale hierarchical structures that dramatically increased surface roughness and hydropho-bicity. The resulting nickel surfaces exhibited water contact angles exceeding 150 degrees, along with excellent corrosion resistance and mechanical durability. This simple, scalable process eliminates the need for toxic chemical coatings or complex surface treatments, illustrating how ECE can enhance material functionality while minimizing environmental impact [13].

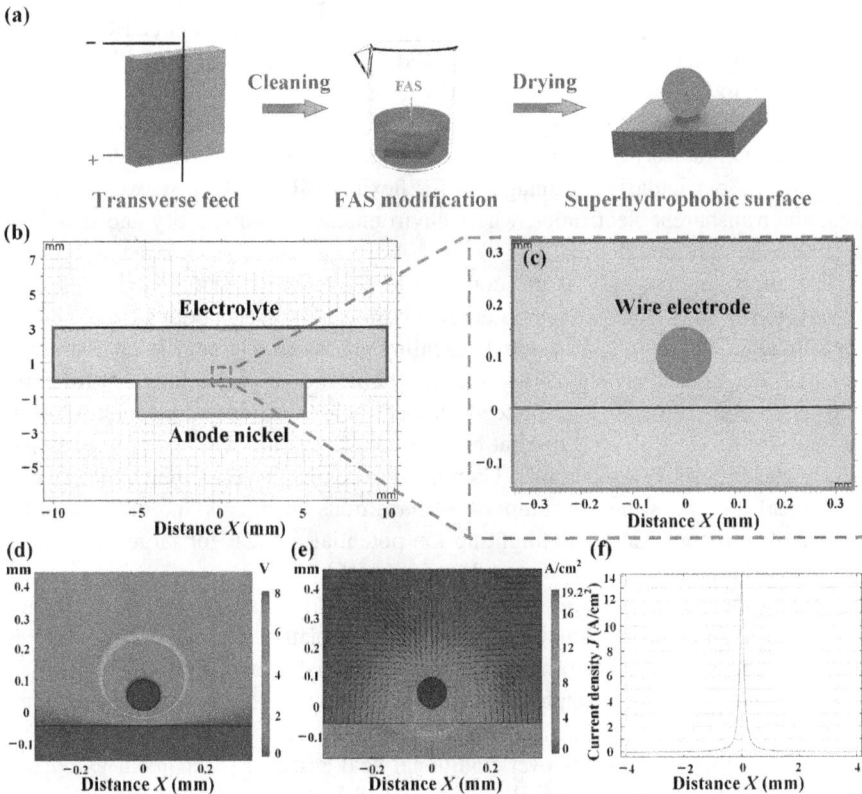

FIGURE 6.2 Wire electrochemical etching (ECE) process and simulation. (a) Flow chart of superhydrophobic nickel sample for wire ECE. (b) Model for wire ECE. (c) Local magnifica-tion for wire ECE. (d) Voltage for wire ECE. (e) Current density for wire ECE. (f) Current density data on the surface of anode nickel plate [13]. (Source: Open access, MDPI Materials.)

The ability of ECE to generate nanostructured porous metal films has also enabled significant advances in electrochemical devices and biosensing technologies. Gryszel et al. demonstrated that nanoporous noble metal films, including gold, platinum, and palladium, can be fabricated through a two-step process involving reactive sputtering of metal oxides followed by electrochemical reduction. The reduction step, driven by ECE, produces highly porous, electrochemically active noble metal structures with substantially increased surface area and capacitance. These films exhibit low impedance and excellent stability, making them ideal for bioelectronic devices, neural interfaces, and electrocatalysis. By avoiding the use of aggressive chemical etchants typically required for dealloying or templating, this approach underscores the compatibility of ECE with green microfabrication strategies for advanced functional materials [14].

ECE has further demonstrated its utility in structuring and patterning conductive oxides, particularly transparent conductive materials such as indium tin oxide (ITO). ITO is widely used in MEMS, sensors, and optoelectronic devices, but conventional wet etching methods often rely on strong acids and suffer from limited precision and poor environmental performance. To overcome these limitations, the Commissariat à l'Énergie Atomique et aux Énergies Alternatives developed a patented ECE method for ITO using microelectrodes and optimized electrolyte formulations. This process enables localized, high-resolution patterning of ITO films without the detrimental undercutting or edge defects typically associated with isotropic chemical etching. The ability to precisely define transparent conductive patterns through electrochemical means is particularly advantageous for flexible MEMS devices, wearable sensors, and transparent electronics, where environmental sustainability and structural precision are paramount [15].

The application of ECE is not confined to conventional metals and oxides but extends to stainless steel, where oxide scale removal represents both a technical and environmental challenge. Traditional descaling processes rely heavily on aggressive acid pickling, generating significant volumes of hazardous waste and harmful emissions. Lee and co-workers introduced an electrochemical descaling method using an ionic liquid–acid electrolyte that enables efficient oxide removal from stainless steel surfaces at room temperature. Compared to conventional methods, this process dramatically reduces the consumption of hazardous chemicals and eliminates the release of nitrogen oxides, highlighting the potential of ECE for large-scale metal finishing with minimal environmental footprint [16].

Beyond metals and oxides, carbon-based conductive materials, particularly graphene, have become integral to MEMS, sensing platforms, and energy devices. Electrochemical approaches provide scalable, controllable, and sustainable routes for producing and modifying graphene and related materials. Pumera et al. comprehensively reviewed electrochemical exfoliation and delamination methods for graphene, emphasizing their advantages over traditional acid-based or mechanical exfoliation techniques. These electrochemical methods yield high-quality graphene flakes and films with fewer structural defects and lower contamination levels. Moreover, electrochemical functionalization enables the tuning of graphene's surface chemistry and electronic properties, further expanding its applicability in sensors, flexible electronics, and energy storage devices. ECE thus plays a central role in advancing sustainable processing of carbon nanomaterials for next-generation microsystems [17].

A critical factor in realizing the full potential of ECE for metals and conductive oxides is the careful optimization of process parameters and the exploration of green electrolyte systems. The addition of sulfuric acid to conventional electrolytes, as demonstrated by Yi et al., enhances isotropic etching behavior and surface quality while reducing reliance on mechanical polishing or hazardous chemicals [12]. Similarly, the use of ionic liquid–acid electrolytes for stainless steel descaling, as reported by Lee et al., illustrates how electrolyte engineering can simultaneously improve process performance and environmental sustainability [16].

Electrochemical modification extends beyond surface structuring to include tailoring of electronic properties in conductive oxides. Zhang and co-workers demonstrated that controlled electrochemical intercalation reactions can significantly alter the conductivity of metal oxides such as titanium dioxide (TiO_2). Their room-temperature, visual intercalation process enables rapid tuning of oxide film properties without the need for high-temperature treatments or toxic dopants. Such electrochemical modification approaches offer green, scalable pathways for engineering functional oxide films compatible with flexible electronics and advanced MEMS applications [18].

While ECE primarily focuses on material removal, it is often employed in conjunction with ECD to enable complex microstructuring and device fabrication. ECD provides a bottom-up approach for building functional coatings, thin films, and nanostructures, while ECE delivers precise, selective material removal. Together, these electrochemical techniques offer a comprehensive toolbox for creating hierarchical micro- and nanostructures with tailored properties. Suchikova et al. conducted a bibliometric analysis of ECE and ECD research, revealing their complementary roles across applications including porous materials, energy devices, biosensors, and photonic structures. This intertwined development underscores the central role of electrochemical processes in enabling scalable, high-precision, and environmentally conscious microfabrication [19].

The continued evolution of ECE for metals and conductive oxides will be driven by advances in green electrolyte systems, real-time process control, and integration with emerging materials platforms. Bio-derived and recyclable electrolytes, along with self-powered ECE systems, offer promising avenues for further reducing environmental impact. The convergence of ECE with advanced materials such as flexible conductive oxides, nanocomposites, and bio-integrated devices will expand its application space and reinforce its role as a cornerstone of sustainable microfabrication. Through these ongoing developments, ECE is poised to play an increasingly vital role in the realization of high-performance, environmentally responsible MEMS and microsystems.

6.4 HYBRID ELECTROCHEMICAL ETCHING: COMBINING ECE WITH PLASMA AND WET ETCHING

ECE has established itself as a highly controllable and environmentally conscious technique for micro- and nanostructuring of semiconductor materials, metals, and conductive oxides. Despite its advantages in material selectivity and reduced chemical consumption, ECE faces intrinsic limitations in terms of etch rate, achievable aspect ratio, and the processing of chemically inert materials. To overcome these challenges, researchers have developed hybrid etching strategies that combine ECE with plasma-based dry etching and conventional wet etching techniques. The integration of these methods harnesses their complementary strengths, enabling the fabrication of complex, multifunctional microstructures with high precision and improved sustainability.

The rationale for hybrid etching lies in the distinct advantages and limitations of each individual process. Plasma etching, particularly reactive ion etching (RIE) and inductively coupled plasma (ICP) techniques, provides excellent anisotropy and depth control but often relies on aggressive gases and high-energy processes. These can lead to surface damage and introduce significant environmental concerns due to greenhouse gas emissions. Wet etching offers simplicity and scalability but typically produces isotropic profiles with limited structural precision. ECE provides highly selective, localized etching with minimal environmental impact but is constrained by relatively slow etch rates and limited capability to achieve deep or vertical structures. By combining these approaches, hybrid etching enables the realization of micro- and nanostructures with enhanced control over geometry, material selectivity, and process sustainability.

A notable example of hybrid etching is its application to SiC, a wide-bandgap semiconductor known for its exceptional hardness, chemical resistance, and thermal stability. These properties make SiC highly attractive for MEMS, power electronics, and harsh-environment sensors, but also render it extremely challenging to etch with conventional methods. Beydoun et al. demonstrated a successful hybrid etching process for SiC that integrates nickel-mask-assisted plasma etching with subsequent ECE. The process begins with the deposition of a thick electroplated nickel mask, which provides the etch resistance required for aggressive ICP/RIE plasma etching using SF_6/O_2 chemistry. This initial step defines deep trenches with good anisotropy, achieving etch depths up to 30 micrometers. To further extend etch depth and introduce layer selectivity, dopant-selective ECE is applied. By exploiting the differential electrochemical behavior of n-type and p-type SiC regions, the electrochemical step selectively corrodes the n-type layers, achieving etch depths exceeding 80 micrometers with superior structural definition. This integrated approach demonstrates how hybrid etching can produce vertical SiC MEMS structures with minimized damage and improved environmental performance compared to purely plasma-based methods [20].

Lithium niobate (LiNbO$_3$, LNO) presents similar challenges for microstructuring due to its chemical inertness and complex crystal structure. As a material widely used in photonic, piezoelectric, and ferroelectric MEMS devices, LNO requires precise, high-quality etching processes that conventional techniques struggle to deliver. Shen and co-workers developed a hybrid etching strategy for LNO that combines dry plasma etching with wet chemical processing. The process begins with Cl-based ICP etching, which provides improved anisotropy and reduced sidewall roughness compared to traditional fluoride-based plasma chemistries. A high-temperature reduction treatment is then applied, followed by selective wet etching to smooth sidewalls and remove byproducts. The combination of these steps enables the fabrication of LNO nanodevices with nearly vertical sidewalls, etch rates exceeding 100 nanometers per minute, and reduced reliance on aggressive chemicals. This approach illustrates how hybrid etching can overcome the limitations of individual methods, providing high-precision microfabrication for otherwise challenging materials [21].

The integration of ECE with plasma and wet etching has also proven essential for structuring GaN and other III-nitride materials, which are central to high-performance optoelectronic devices such as LEDs, lasers, and photonic crystals. GaN's chemical stability and mechanical robustness make precise etching difficult with traditional processes. Hoormann et al. introduced a hybrid etching method for GaN that combines nitrogen ion implantation with subsequent ECE. Ion implantation modifies the doping profile and introduces localized damage that selectively suppresses porosification in targeted regions during ECE. This allows for precise spatial control over the formation of meso- and macroporous regions, enabling the fabrication of photonic devices with engineered refractive indices, such as DBRs, VCSELs, and photonic crystals. The combination of ion implantation, plasma processing, and ECE provides the design flexibility and structural control required for advanced GaN-based microsystems [22].

Further advances in hybrid etching have been demonstrated in the fabrication of III-nitride nanowire devices. Yulianto and co-workers reported a wafer-scale transfer process for GaN nanowire LED arrays that integrates femtosecond laser lift-off, plasma etching, and wet etching techniques. The process begins with the growth of GaN nanoLEDs on sapphire substrates, followed by precise nanowire definition using plasma and wet etching. Femtosecond laser lift-off enables damage-free release of the nanowire arrays, which are then transferred onto flexible metal foils with high yield and structural integrity. This hybrid approach facilitates the fabrication of large-area, flexible optoelectronic devices while minimizing environmental impact, highlighting the potential of integrated physical, plasma, and electrochemical processes for scalable manufacturing of advanced microsystems [23].

The fabrication of metallic MEMS components has also benefited from hybrid etching strategies, particularly in the case of gold. Gold is widely used in MEMS due to its superior electrical conductivity, chemical stability, and biocompatibility, but its resistance to conventional plasma etching has historically restricted its patterning options. Green provided a comprehensive review of gold etching, emphasizing the advantages of combining plasma and wet etching to overcome these limitations. High-density plasma reactors with optimized chlorine chemistries have advanced the dry etching of gold, while wet etching with iodine-based and non-aqueous solutions enables isotropic removal for bulk thinning or planarization. This combination allows for the fabrication of high-aspect-ratio gold microstructures with improved dimensional control, demonstrating the versatility of hybrid etching even for noble metal components [24].

Despite these advances, hybrid ECE presents several technical challenges that must be addressed for broader industrial implementation. Careful optimization is required to ensure compatibility between different etching steps, masking layers, and materials. Matching etch rates and maintaining selectivity across techniques is complex, particularly for multilayer or heterostructured substrates. Sequential etching steps can introduce defects or structural stress, while the need for specialized equipment may increase process complexity and costs.

Nevertheless, the benefits of hybrid etching, including enhanced precision, complex geometry realization, material selectivity, and reduced environmental impact, firmly establish its importance for next-generation MEMS fabrication. The continued evolution of hybrid ECE is expected to focus on dopant- and material-selective etching schemes that enable atomic-level structural control, as well as the development of self-limiting and self-aligned processes to improve reproducibility. Real-time process monitoring and AI-assisted control strategies are likely to enhance process stability and yield. Furthermore, the exploration of bio-derived, recyclable electrolytes and green plasma chemistries will advance the environmental sustainability of hybrid etching.

As MEMS technologies advance toward greater complexity, multifunctionality, and broader material integration, hybrid ECE will play a central role in enabling high-precision, environmentally responsible microfabrication for the next generation of microsystems.

REFERENCES

1. French, P. J., Krijnen, G. J. M., Vollebregt, S., & Mastrangeli, M. (2022). Technology development for MEMS: A tutorial. *IEEE Sensors Journal*, 22(11), 10106–10125.
2. Choi, K., Kim, S. W., Lee, J. H., Chu, B., & Jeong, D. Y. (2024). Eco-friendly glass wet etching for MEMS application: A review. *Journal of the American Ceramic Society*, 107, 6497–6515.
3. Wang, R. C., Zhou, Y. H., Lee, Y. H., Chen, H. C., & Chen, C. E. (2024). TENG-driven single-droplet green electrochemical etching and deposition for chemical sensing applications. *Applied Surface Science Advances*, 23, 100634.
4. Zhao, F., Islam, M. M., & Huang, C. F. (2011). Photoelectrochemical etching to fabricate single-crystal SiC MEMS for harsh environments. *Materials Letters*, 65, 409–412.
5. Hochreiter, A., Groß, F., Möller, M. N., Krieger, M., & Weber, H. B. (2023). Electrochemical etching strategy for shaping monolithic 3D structures from 4H-SiC wafers. *Scientific Reports*, 13, 19086.
6. Han, J. (2014). Lateral electrochemical etching of III-nitride materials for microfabrication. U.S. Patent Application US20140003458A1.
7. Hong, H., Lei, X., Wei, J., Tang, W., Ye, M., Sun, J., Zhang, G., Sarro, P. M., & Liu, Z. (2025). Investigation on fabrication of silicon nanopores using an electrochemical passivation etch-stop strategy. *Microsystems & Nanoengineering*, 11, 128.
8. Roos, M. M., Winkler, A., Nilsen, M., Menzel, S. B., & Strehle, S. (2022). Towards green 3D-microfabrication of bio-MEMS devices using ADEX dry film photoresists. *International Journal of Precision Engineering and Manufacturing-Green Technology*, 9, 43–57.
9. Bassu, M., Strambini, L. M., & Barillaro, G. (2011). Advances in electrochemical micromachining of silicon: Towards MEMS fabrication. *Procedia Engineering*, 25, 1653–1656.
10. Hao, X., He, P., & Li, X. (2022). Selective electrochemical etching of cantilever-type SOI-MEMS devices. *Nano and Precision Engineering*, 5(2), 023003.
11. Ivanov, A., & Mescheder, U. (2011). Dynamic simulation of electrochemical etching of silicon. In COMSOL Conference, Stuttgart.
12. Yi, R., Zhan, Z., & Deng, H. (2022). Isotropic tuning of electrochemical etching for the nanometric finishing of metals. *Nanomanufacturing and Metrology*, 5, 283–296.
13. Wu, B., Yan, D., Lin, J., & Song, J. (2023). Wire electrochemical etching of superhydrophobic nickel surfaces with enhanced corrosion protection. *Materials*, 16, 7472.
14. Gryszel, M., Jakešová, M., Lednický, T., & Głowacki, E. D. (2022). High-capacitance nanoporous noble metal thin films via reduction of sputtered metal oxides. *Advanced Materials Interfaces*, 9, 2101973.

15. Commissariat à l'Énergie Atomique et aux Énergies Alternatives. (2017). Method for etching conductive metal oxide layer using microelectrode. *European Patent* EP2553149B1.

16. Lee, J., Kang, Y., Kim, J. S., Park, J., Lee, J. J., & Kim, B. K. (2020). Electrochemical descaling of metal oxides from stainless steel using an ionic liquid–acid solution. *ACS Omega*, 5, 15709–15714.

17. Ambrosi, A., Chua, C. K., Latiff, N. M., Loo, A. H., Wong, C. H. A., Eng, A. Y. S., Bonanni, A., & Pumera, M. (2016). Graphene and its electrochemistry – An update. *Chemical Society Reviews*, 45, 2458–2493.

18. Zhang, Y., Zhang, X., Pang, Q., & Yan, J. (2023). Control of metal oxides' electronic conductivity through visual intercalation chemical reactions. *Nature Communications*, 14, 6130.

19. Suchikova, Y., Nazarovets, S., & Popov, A. I. (2025). Electrochemical etching vs. electrochemical deposition: A comparative bibliometric analysis. *Electrochem*, 6, 18.

20. Beydoun, N., Lazar, M., & Gassmann, X. (2023). SiC plasma and electrochemical etching for integrated technology processes. *Romanian Journal of Information Science and Technology*, 2023(2), 238–246.

21. Shen, B., Zhu, C., & Tang, Y. (2023). Advanced etching techniques of LiNbO$_3$ nanodevices. *Nanomaterials*, 13(20), 2789.

22. Hoormann, M., Braut, M., Stoll, K., & Strassburg, M. (2024). Electrochemical etching of nitrogen ion-implanted gallium nitride: A route to 3D nanoporous patterning. *Physica Status Solidi B*, 261(4), 2400067.

23. Yulianto, N., Wang, X., & Hsu, H. (2021). Wafer-scale transfer route for top–down III-nitride nanowire LED arrays based on the femtosecond laser lift-off technique. *Microsystems & Nanoengineering*, 7(1), 32.

24. Green, T. A. (2014). Gold etching for microfabrication. *Gold Bulletin*, 47(4), 205–216.

7 Energy-Efficient Plasma Etching Techniques for MEMS

7.1 LOW-POWER PLASMA PROCESSING FOR MEMS

Plasma etching is a dry etching process widely used in microelectromechanical systems (MEMS) fabrication to precisely remove material layers. It involves exposing the material to a high-speed stream of plasma generated from gas mixtures (e.g., fluorine or oxygen-based gases). Plasma consists of ions, radicals, and neutral particles that chemically react with the material, creating volatile byproducts that are removed under vacuum conditions. Figure 7.1 shows the plasma etching system configuration. The substrate is positioned on the bottom electrode that is electrically grounded, and the top electrode is connected to a radio frequency (RF) generator [1]. This technique allows for highly anisotropic etching, enabling the creation of intricate microstructures with excellent pattern fidelity [3, 4]. Low-power plasma processing is the technique used to fabricate MEMS devices using plasma-based etching and deposition processes while limiting power consumption. The main advantages of this technology are:

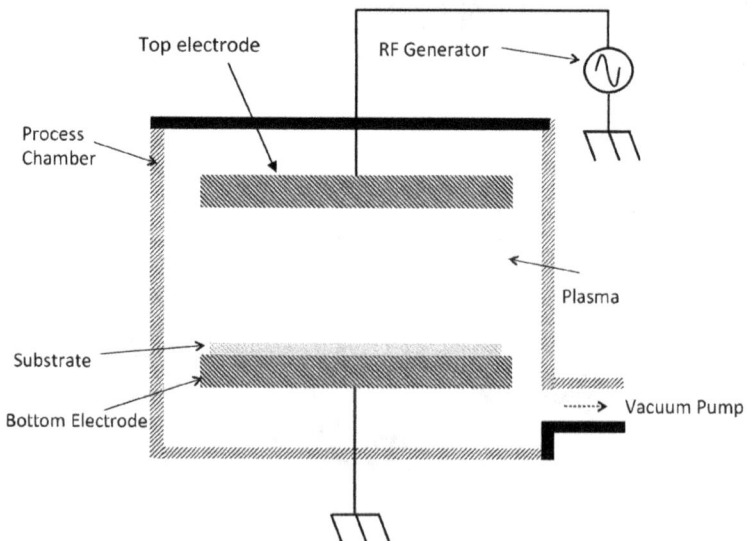

FIGURE 7.1   Illustration of plasma etching process [1].

DOI: 10.1201/9781003674795-7

- Low operating temperature, at which plasma can be generated at temperatures close to room temperature, which makes low-power plasma processing suitable for the treatment of temperature-sensitive materials [5].
- High efficiency, since low-power plasma sources can achieve high electron densities ($>10^{12}$ cm^{-3}) while consuming minimal energy [5, 6].
- Versatility, since this technology can be used for various processes, such as etching, deposition, and surface modification [5].

Low-power plasma processing for MEMS requires precise control of RF power modulation to minimize ion bombardment while maintaining etch efficiency. Table 7.1 lists the comparison between traditional etching and low-power plasma etching [1–6]. There are several techniques that can be used in order to achieve this power optimization.

Pulsed RF power modulation alternates between plasma-active and plasma-off phases to reduce ion energy; it behaves similarly to pulsed light energy sources (i.e., pulsed laser light source) [7, 8]. The active-Glow (plasma-active) phase generates reactive radicals and ions for etching, while the afterglow phase (plasma-off) allows neutral species to dominate, reducing cumulative ion energy exposure. For instance, dual-antenna inductively coupled plasma (ICP) systems demonstrate that pulsed outer coils can shift electron density distributions from center-high to edge-high

---

TABLE 7.1

Comparison between Traditional Etching and Low-Power Plasma Etching [1–6]

Aspect	Traditional Plasma Etching	Low-Power Plasma Etching
Energy Consumption	Higher power levels, typically in the range of hundreds of watts	Lower power levels, often below 100 watts
Etch Rate	Generally faster due to higher plasma density	Slower, but with improved control and precision
Ion Bombardment	Higher energy bombardment, which can cause substrate damage	Reduced ion bombardment energy, minimizing substrate damage
Temperature	Often operates at higher temperatures	Can operate at lower temperatures, suitable for temperature-sensitive substrates
Selectivity	May have lower selectivity due to higher energy processes	Improved selectivity due to gentler etching conditions
Anisotropy	Highly anisotropic, suitable for creating high aspect ratio structures	Less anisotropic, but still capable of directional etching
Material Compatibility	May be limited for some temperature-sensitive materials	Broader material compatibility, including temperature-sensitive substrates
Process Control	Good control, but may require careful tuning to avoid damage	Finer control over etch parameters, allowing for more precise feature creation
Application Range	Widely used for deep etching and high aspect ratio structures	Suitable for delicate structures and thin film etching in MEMS
Cost	Higher operating costs due to increased power consumption	Lower operating costs, but may require longer processing times

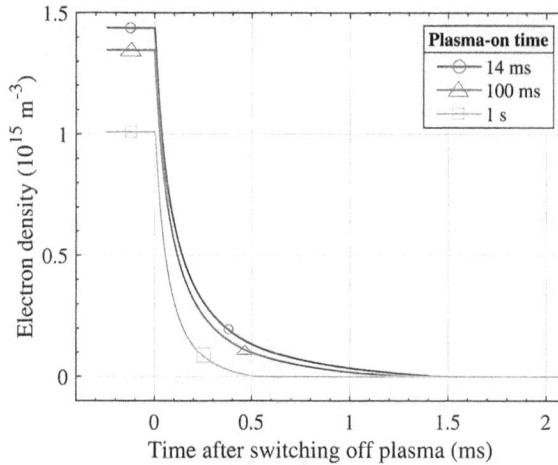

FIGURE 7.2   Time-resolved electron densities during the afterglows of 14 ms, 100 ms, and 1 s plasma-on times [2].

profiles, improving etch uniformity (see Figure 4.2) [7]. It illustrates the configuration of an inductively coupled plasma-reactive ion etching (ICP-RIE) system. In the illustrated system, there are two RF generators, one to create and sustain the plasma and the second to bias the reactants to the substrate [1].

Adjusting the duty cycle percentage, which is the ratio of active plasma time per pulse, directly impacts ion energy distribution. Lower duty cycles (20–40%) reduce high-energy ion bombardment by limiting sustained plasma exposure. In dual-antenna ICPs, increasing the outer coil duty cycle from 20% to 60% shifts time-averaged electron density profiles from center-peaked to edge-peaked, enabling tailored ion flux patterns. While at 50% duty cycle, varying outer coil currents (~5.7–7.7 A) modulate plasma radial uniformity, affecting etch depth consistency. Over-reduction in the duty cycle could lead to a decrease in etching rate, making it a trade-off [7]. For instance, the plot in Figure 7.2 shows the electron density decay during the plasma afterglow. It can be observed in Figure 7.2 that the electron density decays faster for 1 s plasma-on time, compared to 14 ms and 100 ms plasma-on time cases. Also, a lower initial electron density just before the onset of the afterglow phase is observed for the 1 s plasma-on time [2].

It is considered an advanced RF coupling configurations that decouple ion flux from ion energy. ICP-RIE systems use dual RF sources, a 2 MHz power that controls plasma density (ion flux), and 13.56 MHz bias power that regulates ion energy between 50 and 300 eV range [9]. Cryogenic deep reactive ion etching (DRIE) combines RF bias power (50–150 W) with sub −100°C temperatures to achieve vertical profiles with <2° sidewall angle variation [10].

Figure 7.3 illustrates the critical fabrication steps to produce MEMS relays. (a) Fabrication starts with an oriented silicon-on-insulator (SOI) wafer. (b) A thin photoresist is spun atop the SOI wafer and patterned with cantilevered relay designs using standard photolithography. (c) The top Si layer, also known as the device layer, is etched using a cryogenic DRIE process. (d) Release of cantilevers is achieved by

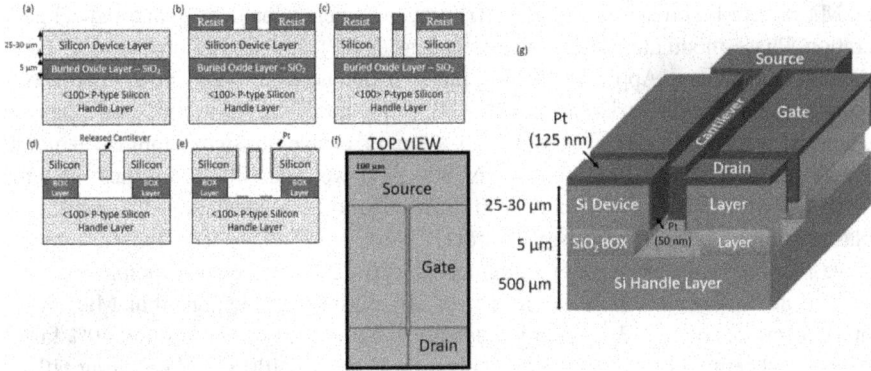

FIGURE 7.3 (a)–(g) Critical fabrication steps to produce MEMS relays [10].

wet etching of the SiO_2 buried oxide (BOX) layer in 49% hydrofluoric (HF) acid. (e) Metallization of sidewalls and surfaces (as contact pads) is performed by RF or DC sputter coating. (f) Top-view optical microscopy image of the fabricated device. (g) 3D schematic of final device [10].

In addition, dual-antenna systems enable spatial control of ionization zones, reducing edge-to-center etch rate disparities by 15–20% [11]. Implementation challenges may occur due to the complex impedance matching for pulsed dual-frequency systems, thermal management during high-power pulses, and process drift monitoring via real-time OES. By integrating these RF modulation strategies, MEMS manufacturers achieve 30–50% reductions in ion energy while maintaining etch rates >3 µm/min for critical structures like comb drives and resonators [10].

ICP systems excel in generating high-density plasmas at lower pressures, critical for MEMS etching and deposition. For instance, Miniaturized ICP designs (e.g., microstrip spiral resonators) reduce power consumption by 30–40% compared to conventional RF sources. The removal of series-wound capacitance in self-resonating circuits improves quality factors and electric field intensity [12]. Dual-antenna configurations enable precise radial plasma uniformity control. Adjusting outer coil currents (5.7–7.7 A) shifts electron density profiles from center-peaked to edge-peaked, reducing etch non-uniformity to <5% [13]. Finally, Helicon plasma modes in ICP systems achieve electron densities >10^{13} cm^{-3}, but require magnetic field tuning to suppress density oscillations. Bi-Maxwellian electron distributions in helicon discharges improve dissociation rates for complex MEMS materials like $LiTaO_3$ [13]. Table 7.2 shows comparison of some key parameters.

TABLE 7.2
Key Parameters for MEMS Optimization

Parameter	ICP Mode	Helicon Mode
Electron density	10^{12}–10^{13} cm^{-3}	10^{13}–10^{14} cm^{-3}
Operating pressure	5–20 mTorr	1–5 mTorr
Etch rate	2–4 µm/min	3–6 µm/min

Microwave-based systems (2.45 GHz) provide superior ion density and radical generation. Phase-modulated surface-wave plasmas (SWP) enhance uniformity in 450 mm diameter chambers. Applying 0°/180° phase differences between input microwaves reduces radial density variations to <8% [14]. TM_{011}-mode microwave plasma-enhanced chemical vapor deposition (MPECVD) systems enable low-temperature diamond/graphene growth for MEMS sensors. At 800 W power and 200 mTorr, these systems achieve plasma densities >10^{12} cm^{-3} with <15% edge-to-center deviation [15]. While cold plasma tuning via electron density control (10^{10}–10^{12} cm^{-3}) allows reconfigurable microwave coupling, reducing power requirements by 50% for photoresist stripping [16].

The advantages of these sources over RF plasmas implement that Microwave plasma sources have 2–3x higher ion density at equivalent power; around 40% faster photoresist removal rates; and lower substrate heating (<100°C). Combining pulsed plasmas with advanced ablation techniques improves MEMS material processing. Nanosecond repetitively pulsed (NRP) discharges generate non-equilibrium plasmas with electron densities >10^{15} cm^{-3} during 5 ns pulses, enabling high-aspect-ratio (HAR) etching without thermal damage. Hybrid laser-plasma ablation models on the other hand, predict 20–30% reduction in nanoparticle contamination during Cu MEMS fabrication. The technique combines target fragmentation kinetics (10^{-9}–10^{-6} s timescales), droplet evaporation control (500–2000 K thermal gradients), and multiphoton ionization for precision material removal [17]. Atmospheric pressure plasma processing (APPP) achieves 31.7 nm RMS surface accuracy using multi-aperture gas flow control, critical for optical MEMS phase plates [18].

There are some implementation challenges connected to this methodology such as material compatibility, since CF_4/Ar gas mixtures require precise tuning (e.g., 28/30 standard cubic centimeters per minute (SCCM) ratio) to maintain >10:1 selectivity for $LiTaO_3$ MEMS resonators [19]. Furthermore, scalability in microwave systems needs custom converters for wafer size transitions (3"→8"), increasing equipment costs. The process tuning for specific MEMS structures presents several challenges that require careful consideration and optimization. Here are the five key challenges elaborated:

1. Different MEMS materials require tailored plasma processes to achieve desired etch profiles and selectivity. For example, silicon nitride MEMS phase shifters require precise tuning of CF_4/Ar gas mixtures (e.g., 28/30 SCCM ratio) to maintain >10:1 selectivity for $LiTaO_3$ MEMS resonators [20]. Furthermore, Dielectrics (e.g., SiO_2, Si_3N_4) require higher ion energies (10–30 eV) for bond breaking compared to silicon [22, 23]. Metals (e.g., Al, Cu) on the other hand, risk redeposition due to low volatility byproducts in fluorine-based plasmas [22]. One of the solutions is that multi-step etching could be done by utilizing low-energy ion bombardment (5–15 eV) for selectivity and high-energy pulses (20–30 eV) for dielectric breakthrough [22]. In addition, combining ICP for radical generation with capacitively coupled plasma (CCP) for controlled ion acceleration [23].

2. As MEMS structures become smaller, plasma processes must be fine-tuned to maintain critical dimensions (CD) and avoid aspect-ratio dependent etching effects. Sub-μm MEMS structures, for instance, demand <5% duty cycle variations to prevent aspect-ratio dependent etching [21]. Etch rates decrease

exponentially in HAR features (>50:1) due to ion shadowing and neutral transport limitations [22, 23]. Cryogenic etching could be used to stabilize passivation layers, enabling vertical profiles in 100 nm wide features. Furthermore, conformal doping could be used to pre-treat sidewalls with boron or phosphorus to create etch-resistant layers in silicon HAR structures [23].

3. Low-power plasma processes must be optimized to minimize induced stress in MEMS structures, which can affect device performance. Cryogenic DRIE combines RF bias power (50–150 W) with sub-−100°C temperatures to achieve vertical profiles with <2° sidewall angle variation, helping manage stress in etched structures [21]. Using laser interferometry to measure wafer curvature during deposition, adjusting RF power to maintain <50 MPa stress. In addition, Annealing protocols could be implemented to perform 400°C post-etch annealing in N_2 to relieve stress in silicon springs without damaging complementary metal–oxide–semiconductor (CMOS) layers [23]. Furthermore, thermal shunt designs could be utilized to dissipate plasma heat when applied near sensitive structures.

4. New low-power plasma techniques must be compatible with established MEMS fabrication workflows and equipment. For instance, the combination of commercially available CMOS processes from foundries with subsequent processing in dedicated MEMS facilities can help address integration challenges. On the contrary, thermal budget conflicts arise when combining low-power plasmas (≤300°C) with high-temperature metallization steps (>400°C). Hence, using plasma-activated wafer bonding at 250°C with ≤5 nm RMS surface roughness; patterning SiO_2 etch to stop before bonding to enable DRIE release without wet processing; and selecting carrier wafers with coefficient of thermal expansion (CTE) ≤ 3 ppm/°C (e.g., fused silica) to minimize thermal mismatch [23] can help overcome this integration challenge.

5. Implementing real-time monitoring and feedback systems to maintain consistent low-power plasma performance across wafer batches is crucial, for pulsed dual-frequency systems, complex impedance matching, and thermal management during high-power pulses are necessary. In addition, developing well-defined pass/fail criteria for each process step that can be easily inspected using common metrology equipment is essential for quality control [21]. Furthermore, AI-driven predictive modeling is a huge advantage to forecast etch profiles with <5% error by training neural networks on plasma parameters (e.g., 13.56–60 MHz RF, 10–100 mTorr pressure). Integrating X-ray photoelectron spectroscopy to monitor surface chemistry at 1 nm resolution during etching would be helpful in terms of etching in CD (<100 nm) [22].

7.2 PULSED PLASMA ETCHING FOR HIGH-SELECTIVITY MEMS FABRICATION

In the development of MEMS, pulsed plasma etching (PPE) has been an innovative advancement with regard to selectivity, anisotropy, and damage mitigation—properties crucial to HAR device structures fabrication. PPE converts static plasma process into a dynamically changing one in the time domain—as it usually operates at kHz-scale with duty cycles from 10% to well over 80%. The modulation enables process engineers to partly decouple physical sputtering and chemical reactions from each other, optimize ion energy deposition as well as improve etch anisotropy. Pulsed plasma discharges, in contrast to continuous wave (CW) are modulated RF power through the source, bias, or both, in the range of 100 kHz to 13.56 MHz [24, 25].

Pulsing provides "off" times which impacts plasma kinetics and the sheath ensuring potential for the ion energy distribution functions (IEDF), electron temperature (T_e), and neutral-to-ion flux ratios to be controlled and decrease damage and profile in high-density, HAR MEMS fabrication. Kim and Yeom (2023) demonstrated that asynchronous pulsing of RF source and bias power can lead to enhanced profile verticality against aspect-ratio-dependent etching (ARDE) in nanoscale Si trench etching, whereas studies have further shown that PPE also mitigates charging and decreases defectivity, important for complex MEMS device geometries [25].

One major benefit of PPE is its ability to reduce substrate sputter damage and reduce substrate heating. In CW plasmas, this is exacerbated by continuous ion impact and vacuum ultraviolet (VUV) photon flux leading to heat generation. Duty cycle is the time an RF waveform is on average, and pulsed operation, typically 10–50%, enables ion-neutral recombination in the off-state to reduce the ion energy to <20 eV, which is well below most sputter thresholds, e.g., SiO_2, and photoresist layers. While Kim and Yeom (2023) mainly discussed asynchronous pulsing in particular for etch profile control, but the concept of offering off-periods in plasma operation as an intrinsic advantage for passively cooling when dealing with thermally sensitive layers such as polymers, metal interconnects, or low-k dielectrics is generally accepted [25].

PPE allows ion energy distribution, which reduces physical damage during the etching process. According to Economou (2014), bias pulsing in electronegative plasmas such as Cl_2 can substantially mitigate micro-trenching and other profile anomalies that sub-100 nm features by pulsing ion energy. This time-based control of the RF substrate bias is a technique that enables accurate control of etch directionality, which reduces substrate damage necessary for advanced device manufacturing [24]. Likewise, Agarwal et al. (2008) also computationally investigated pulsed multi-frequency CCP systems in high aspect-ratio features for dielectric etching [26]. They found that in electronegative gas environments, pulsing of both source and bias frequencies enables the extraction of negative ions during plasma afterglow leading to reduced charging on the sidewall. The mechanisms therefore alleviate such profile distortions like twisting and bowing in just sub-10 nm wide trenches.

Standard ICP systems ion energies usually exceed the range of 100 eV, which allows the atoms to be displaced and perform surface amorphization or roughening. Miyazoe et al. (2017) presented a highly selective dry etch process for block copolymers with gas-pulsing CO/H_2 plasmas [27]. They managed to get rid of the poly(methyl methacrylate) (PMMA, $-[CH_2-C(CH_3)(COOCH_3)]-n$) completely with cycle-by-cycle alternation between deposition and etch, demonstrating etch selectivity well beyond conventional O_2 plasma transfer methods. It was non-destructive on the polymer backbone, which potentially makes this one of the "safest" DSA pattern transfer methods.

When it comes to HAR MEMS structuring, PPE delivers superior anisotropy and selectivity compared to conventional methods. MEMS devices often demand trench depths >50 μm with CD <2 μm, corresponding to aspect ratios exceeding 25:1 [28]. Conventional etching struggles to achieve vertical sidewalls and consistent etch depths across dense and sparse pattern regions due to ARDE.

These features demand precise control of ion angular distribution and sidewall passivation. Using a 13.56 MHz asynchronous PPE signal with substrate bias duty cycles ranging from 10% to 50% using Cl_2/Ar, Kim and Yeom (2023) showed that etching of silicon controls ion dose delivery, preserving the vertical profiles in 100 nm-wide trenches [25]. Near-vertical profiles with less than 5° deviation were achieved using C_4F_8 in the plasma due to enhanced fluorocarbon sidewall passivation. The downside of this was the reduced Si-to-mask selectivity from ~15:1 (with CF_4) to ~5:1 caused by increased polymer deposition. Williams et al. (2003) demonstrated deep trench etching using PPE in an ICP system, achieving aspect ratios over 100:1 for trenches around 1 μm wide and more than 100 μm deep [29]. By using synchronized pulsing, they were able to reduce sidewall scalloping and achieve trench sidewall angles within 1° of vertical.

Wang et al. (2018) developed a coupled 2D plasma and 3D feature-scale simulation framework to investigate silicon etching using a three-step cyclic pulsed ICP process consisting of oxidation, etching, and cleaning [30]. By synchronizing RF bias (2 MHz) and ICP source pulsing at 2 kHz with a 30% duty cycle and zero phase shift, they achieved improved vertical trench profiles and enhanced Si:mask etch selectivity—approaching 12:1—while mitigating profile distortion effects associated with ARDE. This approach is particularly relevant for HAR structures in advanced device and MEMS fabrication.

Williams et al. (2003) demonstrated the potential of PPE with SF_6/C_4F_8 chemistries to achieve high etch selectivity and smooth sidewalls when patterning silicon, emphasizing the effectiveness of synchronized pulsing in minimizing scalloping and preserving vertical profiles [27]. Yamakawa et al. (2004) showed oxide etching using non-equilibrium microwave-excited CF_4/He plasmas under atmospheric pressure [31]. This was possible through pulsed microwave excitation at 2.45 GHz, which helped them achieve etch rates over 3 μm/min and SiO_2: Si selectivity as high as 200:1. Bliznetsov et al. (2013) etched AlN films over Mo electrodes in MEMS resonator applications by coming up with a Cl_2/BCl_3/Ar-based ICP-RIE process [32]. They achieved surface roughness as low as 1.78 nm RMS and AlN:Mo selectivity of up to 4:1 by careful parameter tuning and buffer gases, while maintained reasonable etch rates and suppressed microtrenching.

PPE is challenging due to the challenges associated with it in the process complexity and frequency control domain. Bias and source power must be synchronized, as misalignment can lead to non-uniform etch rates or plasma extinction. For example, extinction is more probable in electronegative gases such as SF_6 or Cl_2 since the off-phase would enable electron attachment, making reignition difficult. Moreover, frequency selection is another problem since frequencies above 100 kHz would be similar to CW power and those below 1 kHz would lead to excessive passivation or incomplete etching due to radical depletion or redeposition.

By maintaining short off-times and operating at controlled pressures, reliable plasma reignition is possible, and electron attachment losses are minimized as reported by Yang et al. (2015) for Cl_2-based electronegative plasmas [33]. In nanoscale etching processes where charging and process uniformity are sensitive to plasma stability, these factors are critical to sustaining stable pulsed discharges. An optimized UV nanosecond laser micromachining process for SiC MEMS accelerometers was reported by Shi et al. (2018) using 35 kHz pulse repetition frequency and four processing laps for reduced thermal damage and clean etch profiles [34]. However, asynchronous pulsing introduces up to 3–5 control parameters—bias delay, rise/fall time, duty cycle, frequency—that exponentially increase system calibration complexity [24].

Modern systems incorporate high-resolution pulse controllers and in-situ diagnostics such as optical emission spectroscopy (OES) and Langmuir probes to tackle these challenges. Lisovskiy et al (2007) synchronized a spectrometer and RF source using pulse control signals to address the complexities of endpoint detection during PPE [35]. This led to improved reliability of endpoint identification during dynamic etching processes due to coordinated inspection and plasma modulation. In addition, research is being carried out on adaptive pulse sequencing algorithms where machine learning (ML) models modify pulse parameters based on near real-time sensor feedback leading to the repeatability of high-volume MEMS manufacturing.

PPE has been employed in a variety of MEMS applications. PPE protect the vertical profiles necessary for resonance stability during etching of trenches up to 100 μm deep and ≤ 2 μm wide in deep trench resonators [36]. In RF MEMS based on AlN, PPE provided correct pattern transfer without dielectric or piezoelectric properties defects [32]. The femtosecond laser micromachining with ICP plasma etching for SiC by Wu et al. (2020) showed smoother surfaces and no carbon residues [37].

In conclusion, PPE is a catalyst technology for next-generation MEMS manufacturing that provides unique selectivity, profile control, and damage mitigation in the fabrication of HAR features. It outperforms steady CW plasma techniques, particularly for fragile or composite substrates, by reducing ion energy flux and suppressing microtrenching as well as enabling radiation control. However, unlocking its full industrial potential rests with the ability to control multi-variable pulsing processes. Go-to future directions are tightly closed-loop control systems for real-time plasma diagnostics and hybrid plasma modalities providing pulsing precision along with atomic layer etch features.

7.3 AI-OPTIMIZED PLASMA CONTROL FOR GREEN MEMS MANUFACTURING

The integration of artificial intelligence (AI) into MEMS fabrication marks a paradigm shift in plasma process control. Conventional plasma etching techniques rely on static recipes and operator-defined parameters, often leading to suboptimal energy usage, material waste, and inconsistent results. In contrast, AI-optimized systems enable real-time adjustments to plasma conditions based on sensor feedback, predictive modeling, and ML algorithms. These systems offer increased uniformity, higher yield, and reduced power consumption—key goals for sustainable, large-scale MEMS manufacturing.

ML, especially deep learning (DL), is increasingly used in plasma process optimization to reduce energy consumption and improve etching outcomes. Traditional rule-based control methods are limited in handling the complex, nonlinear behavior of plasma systems. In contrast, ML models trained on simulation or experimental data can predict process outcomes and adjust parameters in real-time.

DL-based surrogate model trained on data from hybrid plasma equipment simulations predicted spatial plasma characteristics—such as ion flux and electron temperature—with high accuracy [34]. When combined with multi-objective particle swarm optimization (MOPSO), the framework enabled rapid identification of optimal etching conditions that maximize uniformity and etch rate while minimizing energy input.

ML has broader potential in plasma modeling, diagnostics, and control. Supervised learning techniques—like neural networks and support vector machines—can model plasma–surface interactions directly from data, offering efficient alternatives to physics-based approaches [35]. These surrogate models are especially useful when theoretical understanding is limited, or simulations are computationally expensive.

These advances show that ML enables both forward prediction and inverse process design, paving the way for energy-efficient, AI-driven plasma control systems in MEMS manufacturing.

Real-time optimization plays a vital role in modern plasma etching systems for MEMS fabrication. By integrating sensor feedback and control algorithms, these systems can dynamically adjust etching parameters—such as RF power, gas flow rate, or bias voltage—to maintain energy efficiency and etching precision. This active feedback loop minimizes overprocessing, reduces energy spikes, and ensures consistent process outcomes across wafers.

Fusion plasma research offers relevant insights into real-time plasma control. For instance, real-time optimization techniques in the RFX-mod system (Reversed Field eXperiment—modified) utilize sensor-actuator decoupling and lightweight matrix operations to suppress harmonic distortion [36]. Although originally developed for magnetic field stabilization, these techniques can be adapted to microfabrication by enabling rapid plasma condition adjustments to prevent over-etching or surface damage.

Similarly, an optimal real-time controller for plasma position stabilization in Tokamak à configuration variable (TCV) involves reduced-order modeling and predictive control strategies that minimize deviation under rapid plasma fluctuations [37]. In MEMS etching, similar strategies can predict and compensate for instabilities in ion energy delivery or chamber pressure changes, improving yield and reducing power waste.

Incorporating these strategies into MEMS-scale etching processes—via AI or real-time control hardware—offers the potential for more stable, sustainable, and autonomous fabrication systems. The integration of predictive maintenance (Pd.M.) powered by AI is becoming a critical element in achieving energy-efficient and fault-tolerant MEMS plasma etching systems. Traditional maintenance strategies—whether reactive or schedule-based—often result in excessive downtime or unnecessary interventions. AI-driven Pd.M. offers a more sustainable alternative by analyzing system behavior in real time, identifying anomalies, and forecasting component failures before they occur.

Recent studies underscore the potential of AI models, particularly neural networks and DL algorithms, in enhancing diagnostic accuracy and maintenance decision-making. These models are capable of processing complex data patterns generated during plasma etching—such as temperature fluctuations, pressure variations, and RF signal irregularities—to detect subtle deviations indicative of tool wear, chamber contamination, or process drift [38].

The use of edge computing further enhances this capability by enabling localized, real-time monitoring and analysis, reducing latency and bandwidth demands. For MEMS plasma systems, this translates to early fault detection and reduced energy waste associated with reprocessing or equipment downtime. Moreover, adaptive AI systems can evolve with new process data, continually refining their predictive models and diagnostics capabilities without manual retraining.

A case study from the computing continuum domain illustrated how dynamic DL and federated learning approaches improved system stability and reduced operational costs through proactive maintenance scheduling and component health assessments [38]. These principles can be directly applied to MEMS manufacturing environments, where unplanned system faults can jeopardize etching uniformity and yield.

The convergence of AI, predictive diagnostics, and real-time data analytics thus offers a robust framework for building more resilient, energy-efficient MEMS fabrication systems—shifting maintenance from a cost center to a value generator. Despite their advantages, AI-based plasma systems come with implementation challenges. Training robust models requires access to large, high-quality datasets that may not be readily available across different fabrication environments. Additionally, integrating AI with legacy etching tools and ensuring cybersecurity in data-driven manufacturing pose nontrivial barriers.

Looking ahead, federated learning (training models across decentralized data sources) and hybrid physical-data-driven models could overcome these limitations, making AI plasma control more scalable and secure. Combining AI with digital twin simulations also promises improved accuracy in virtual process prototyping before actual runs, thus reducing both trial-and-error cycles and associated energy costs.

7.4 REACTIVE ION ETCHING PROCESS OPTIMIZATION FOR LOW ENERGY CONSUMPTION

Reactive ion etching (RIE) processes can be optimized to reduce energy consumption while preserving the structural integrity of sensitive MEMS devices. One effective approach involves minimizing the RF power applied during etching. For example, reducing the RF bias to 10 W in HBr-based RIE significantly improved the etch selectivity between silicon and polymer masks, reaching a ratio of 36.1. This low-energy regime promoted chemical etching over physical sputtering, as evidenced by the exponential increase in selectivity and corresponding decrease in etch rate at lower RF power. The trade-off between selectivity and speed is critical for applications requiring HAR nanostructures with smoother profiles and fewer defects [39].

Another strategy involves operating at low RF power and chamber pressure. In one study, using CF_4, SF_6, and O_2 gases at 30 W and 2 Pa yielded vertical and smooth-walled silicon features. This setup suppressed physical sputtering and helped avoid roughness and defects typically found in Bosch-type processes. Such methods are particularly valuable for damage-sensitive MEMS and microfluidic devices [40].

Cryogenic RIE, conducted at substrate temperatures near −120°C using SF_6/O_2 plasma, also contributes to low-energy, damage-controlled etching. At these temperatures, sidewall passivation is enhanced through the formation of protective SiO_xF_y films, which reduce trenching and ion impact. This technique is especially beneficial for delicate or porous MEMS structures that require precision without compromising structural stability [41].

For oxide-based MEMS, low-power RIE using He, Ar, and fluorocarbon gases has demonstrated reduced substrate degradation. Processes operated around 0.05 W/ cm^3 with tuned gas mixtures and RF bias frequencies yielded smoother profiles and improved chemical selectivity, further highlighting how gas control impacts low-damage results [42].

In the case of piezoelectric MEMS, such as those based on $LiNbO_3$, a pulsed Ar/ SF_6 RIE process was applied to achieve deep etching without damaging the active material. By optimizing the pulse duty cycle and minimizing electrical and thermal stress, the process successfully preserved piezoelectric performance while maintaining etch precision [43].

The adoption of ICP systems also plays a critical role in enabling low-energy RIE. Unlike CCP sources, ICP reactors allow separate control of ion energy and plasma density. This decoupling lowers ion bombardment while maintaining high plasma density for effective etching. ICP-based systems are particularly advantageous for handling sensitive materials like InP and GaN in MEMS and photonic devices [44].

Recent innovations in pulsed plasma systems have further advanced energy-efficient etching. Pulsing the plasma between 1 and 10 kHz with duty cycles of 20–80% leads to a reduction in ion energy and improved etch selectivity. These systems offer a promising solution for minimizing damage while retaining profile control and anisotropy [45]. In MEMS processing of $LiNbO_3$ structures, a pulsed Ar/SF_6 RIE method was used to achieve deep etches with minimal material degradation. Pulsing the plasma allowed thermal relaxation between cycles, which improved pattern fidelity and prevented overheating of piezoelectric layers [43].

Reducing plasma-induced defects is essential for achieving reliable MEMS devices through RIE. One widely used refinement involves controlling the RF bias power to limit ion bombardment. For example, reducing the RF bias to 10 W in HBr-based RIE promotes chemical etching and reduces surface damage on the silicon substrate [39]. Cryogenic etching also helps address common RIE-induced defects. When operated around −100°C, the SF_6/O_2 plasma forms thin passivation layers on etched surfaces. These layers improve sidewall protection and significantly reduce trenching and micromasking effects, leading to smoother profiles [41]. This technique is particularly valuable for preserving the integrity of advanced CMOS and MEMS devices that demand precise feature control [46].

Etch profile control plays a key role in defect suppression. By optimizing gas ratios—such as increasing C_4F_8 content in $CF_4 + C_4F_8$ plasma—sidewall passivation is enhanced, reducing notching and bowing. The addition of helium helps equalize ion flux, mitigating localized heating and associated deformation [42].

Another source of defects is heat buildup during etching, especially on large-area wafers. Karouta (2014) proposed process improvements such as backside helium cooling and the use of thermal paste to stabilize wafer temperature. These steps improved thermal contact and prevented pattern distortion and delamination during long etch runs. The study also emphasized pressure and gas control to manage etch lag and micro-masking, common sources of surface non-uniformities [44].

Charging effects, especially in dielectric layers, can also introduce pattern distortion or etch stop issues. Increasing process pressure reduces the mean free path of ions, resulting in less energetic ion collisions at the wafer surface. This adjustment softens ion impact and reduces the risk of charge-related damage. Such refinements are often applied in deep trench and low-k dielectric applications [47].

Atomic layer etching (ALE) has become a key tool for minimizing plasma-induced damage. This process uses cyclic steps involving surface modification and low-energy ion removal to ensure layer-by-layer etching precision. ALE enables near-atomically smooth profiles and drastically reduces sidewall roughness and line edge defects. When implemented in dual-frequency CCP reactors, ALE provides an ideal balance between etch control and plasma-induced defect minimization [45].

A practical industrial case study of low-power RIE adoption is demonstrated in the fabrication of a CMOS-MEMS flow sensor using the UMC 0.18 μm process. The device was manufactured through a commercial foundry process involving anisotropic dry etching of silicon oxide layers using a CHF_3/O_2 plasma mixture. This step was optimized to ensure effective material removal while minimizing ion bombardment damage and maintaining compatibility with standard CMOS flows. The post-processing flow includes the RIE oxide etch and XeF_2 silicon release steps, highlighting the integration of low-energy dry etching in the fabrication of suspended MEMS structures. The process was designed to maintain a low thermal budget and reduce physical stress during fabrication, making it suitable for sensitive MEMS applications. This industrial implementation illustrates how low-power etching can be successfully incorporated into a standard CMOS infrastructure without significant changes to existing manufacturing protocols [48].

Another example of industrial-level advancement in low-power RIE is presented in the development of a real-time feedback control system for the Applied Materials 8300 Hexode Reactive Ion Etcher. This system implemented closed-loop plasma monitoring to stabilize key parameters such as ion energy and fluorine concentration during etching. Etch rate behavior was compared under open-loop and closed-loop control across four real-world disturbances: (a) wall moisture variation, (b) an O_2 leak disturbance, (c) wafer loading effects, and (d) low initial wall moisture. In all cases, the feedback system significantly reduced etch rate drift and variability, improving consistency and process robustness over time. These improvements not only enhanced etch quality and repeatability but also enabled more energy-efficient operation by maintaining optimal plasma conditions throughout each run. This study demonstrates how adaptive control strategies can effectively support low-power RIE adoption in industrial semiconductor manufacturing [49].

While both industrial case studies demonstrate the feasibility of low-power RIE in practical applications, their scopes differ significantly. The UMC CMOS-MEMS sensor case [48] showcases seamless integration of low-energy etching in a standard foundry, highlighting compatibility and process efficiency in commercial MEMS. In contrast, the feedback-controlled RIE system [49] focuses on tool-level optimization and real-time plasma stabilization, addressing variability and power efficiency at the equipment level. Together, these examples illustrate that low-power RIE adoption can be achieved both through advanced tool control and through careful process integration, suggesting that a hybrid strategy may offer the most robust path forward in scalable, sustainable MEMS production.

REFERENCES

1. Huff, M. (2021). Recent advances in reactive ion etching and applications of high-aspect-ratio microfabrication. *Micromachines*, 12, 991.
2. Hasani, M., Donders, T. J. M., & Beckers, J. (2023). Temporal afterglow between two pulses of repetitively pulsed argon-acetylene plasma: Measuring electron and negatively charged species densities. *Journal of Physics D: Applied Physics*, 56, 365204.
3. Rahman, H. U. (2012). Plasma based dry release of MEMS devices. In *InTech*. https://doi.org/10.5772/28420
4. Lucibello, A., Proietti, E., Catoni, S., Frenguelli, L., Marcelli, R., & Bartolucci, G. (2007). RF MEMS switches supported by polymeric structures. *International Semiconductor Conference*, 1, 259–262.
5. Galchev, T., Welch, W., & Najafi, K. (2007). Low-temperature MEMS process using plasma activated silicon-on-silicon bonding. *IEEE International Conference on Micro Electro Mechanical Systems*, 309–312.
6. Iza, F., & Hopwood, J. A. (2003). Low-power microwave plasma source based on a microstrip split-ring resonator. *IEEE Transactions on Plasma Science*, 31, 782.
7. Lu, C., Huang, J., Zhang, Y., Gao, F., & Wang, Y. (2023). Modulation of the plasma radial uniformity in pulsed dual-antenna inductively coupled plasmas. *Physics of Plasmas*, 30(6), 063506.
8. Hao, Q., Kim, P., Nam, S. K., Kang, S., & Donnelly, V. M. (2023). Real-time monitoring of atomic layer etching in Cl_2/Ar pulsed gas, pulsed power plasmas by optical emission spectroscopy. *Journal of Vacuum Science & Technology A 41, 032605*
9. Lishan, D., PlasmaTherm. (.). Plasma etching: Comparing PE, RIE and ICP-RIE. https://www.corial.com/, White paper
10. Horstmann, B., Pate, D. S., Smith, B., Mamun, M. A., Atkinson, G., Özgür, Ü., & Avrutin, V. (2024). Cryogenic DRIE processes for high-precision silicon etching in MEMS applications. *Journal of Micromechanics and Microengineering*, 34, 075008.
11. Lu, C., Huang, J., Zhang, Y., Gao, F., & Wang, Y. (2023). Modulation of plasma radial uniformity in pulsed dual-antenna inductively coupled plasmas. *Physics of Plasmas*, 30(6), 063506.
12. Zhou, B., Liao, B., & Zhu, S. (2006). Micro-strip spiral resonator in miniaturized inductively coupled plasma source. *8th International Conference on Solid-State and Integrated Circuit Technology*, 560–562.
13. Zhang, T., Jiang, K., Liu, Z., Yang, L., Zhang, H., Ouyang, J., & Chen, Q. (2020). Characteristics of inductively coupled plasma (ICP) and helicon plasma in a single-loop antenna. *Plasma Science and Technology*, 22, 085405.
14. Kim, D., Kang, K., Shon, C., Kim, J., & Han, S. (2024). Uniformity enhancement of a microwave surface-wave plasma by a field agitation. *Japanese Journal of Applied Physics*, 63, 026001.

15. Jiang, Y., Aranganadin, K., Hsu, H., Lin, M., & Yen, J. (2019). Design and modeling of a microwave PECVD system at 2.45 GHz. *2019 International Vacuum Electronics Conference (IVEC)*, 1–2.
16. Krasik, Y. E., Leopold, J. G., Shafir, G., Cao, Y., Bliokh, Y. P., Rostov, V. V., Godyak, V., Siman-Tov, M., Gad, R., Fisher, A., Bernshtam, V., Gleizer, S., Zolotukhin, D., & Slutsker, Y. (2019). Experiments designed to study the non-linear transition of high-power microwaves through plasmas and gases. *Plasma*, 2(1), 51–64.
17. Heya, M., Furukawa, H., Tsuyama, M., & Nakano, H. (2021). Simulations of the effects of laser wavelength, pulse duration, and power density on plume pressure generation using a one-dimensional simulation code for laser shock processing. *Journal of Applied Physics*, 129(23), 235108.
18. Li, D., Li, N., Su, X., Liu, K., Ji, P., & Wang, B. (2019). Continuous phase plate structuring by multi-aperture atmospheric pressure plasma processing. *Micromachines*, 10(4), 260.
19. Majd, Y., Castro, J. M., Mansoorzare, H., & Abdolvand, R. (2023). Characterizing ICP etching of lithium tantalate for micro-acoustic applications. *IEEE International Conference on Micro Electro Mechanical Systems*, 586–589.
20. McNulty, K., van Niekerk, M., Deenadayalan, V., Errando-Herranz, C., Pruessner, M. W., Fanto, M., & Preble, S. F. (2022). Wafer-scale fabrication of silicon nitride MEMS phase shifters with XeF_2 release. SPIE *OPTO.*, Proceedings, 12006, 32.
21. Beghini, M., Grossi, T., Macoretta, G., Monelli, B. D., Senegaglia, I., del Turco, P., Fardelli, A., & Morante, F. (2024). Tuning modal behaviour of additively manufactured lattice structures. *Journal of Engineering for Gas Turbines and Power*, 146(7), 071008.
22. Graves, D. B., Labelle, C. B., Kushner, M. J., Aydil, E. S., Donnelly, V. M., Chang, J. P., Mayer, P., Overzet, L., Shannon, S., Rauf, S., & Ruzic, D. N. (2024). Science challenges and research opportunities for plasma applications in microelectronics. *Journal of Vacuum Science & Technology B*, 42(4), 042202.
23. Galchev, T. V., Rowe, A., & Najafi, K. (2011). A low-temperature high-aspect-ratio MEMS process using plasma activated wafer bonding. *Journal of Micromechanics and Microengineering*, 21(4), 045020.
24. Economou, D. J. (2014). Pulsed plasma etching for semiconductor manufacturing. *Journal of Physics D: Applied Physics*, 47(30), 303001.
25. Kim, H. J., & Yeom, G. Y. (2023). Asynchronously pulsed plasma for high aspect ratio nanoscale Si trench etch process. *ACS Applied Nano Materials*, 6(11), 10097–10105.
26. Agarwal, A., Stout, P., & Rauf, S. (2008). Characteristics of pulsed capacitively coupled plasma for plasma etching. *Bulletin of the American Physical Society*, 53(10).
27. Miyazoe, H., Jagtiani, A. V., Tsai, H. Y., Engelmann, S. U., & Joseph, E. A. (2017). Highly selective dry etching of PS-PMMA block copolymer by gas pulsing CO-based plasmas. *Journal of Physics D: Applied Physics*, 50(20), 204001.
28. Chaudhuri, A. R., Heli, P., Severi, S., Van Hoof, R., Du Bois, B., Tilmans, H., Francis, L. A., & Witvrouw, A. (2013). Ultra-narrow, high aspect ratio trenches for use in miniaturized poly-SiGe accelerometers. *ECS Meeting, Meet. Abstr., MA2013-02*, 2646, 1–2.
29. Williams, K. R., Gupta, K., & Wasilik, M. (2003). Etch rates for micromachining processing—Part II. *Journal of Microelectromechanical Systems*, 12(6), 761–778.
30. Wang, J.-C., Tian, W., Rauf, S., Sadighi, S., Kenney, J., Stout, P., Vidyarthi, V. S., Guo, J., Zhou, T., Delfin, K., Lundy, N., Pandey, S. C., Guo, S., & Sandhu, G. S. (2018). A model for etching of three-dimensional high aspect ratio silicon structures in pulsed inductively coupled plasmas. *Plasma Sources Science and Technology*, 27(9), 094003.

31. Yamakawa, K., Hori, M., Goto, T., Den, S., Katagiri, T., & Kano, H. (2004). Ultrahigh-speed etching of SiO_2 with ultrahigh selectivity over Si in microwave-excited non-equilibrium atmospheric pressure plasma. *Applied Physics Letters*, 85(4), 549–551.

32. Bliznetsov, V., Johari, B. H. B., Chentir, M. T., Li, W. H., Wong, L. Y., Merugu, S., Zhang, X. L., & Singh, N. (2013). Improving aluminum nitride plasma etch process for MEMS applications. *Journal of Micromechanics and Microengineering*, 23(11), 117001.

33. Yang, K. C., Park, S. W., Shin, T. H., & Yeom, G. Y. (2015). Application of pulsed plasmas for nanoscale etching of semiconductor devices: A review. *Journal of the Korean Institute of Surface Engineering*, 48(5), 360–370.

34. Shi, Y., Sun, Y., Liu, J., Tang, J., Li, J., Ma, Z., Cao, H., Zhao, R., Kou, Z., Huang, K., Gao, J., & Hou, T. (2018). UV nanosecond laser machining and characterization for SiC MEMS sensor application. *Sensors and Actuators A: Physical*, 276, 196–204.

35. Jeong, Y. S., Hwang, S., & Ko, Y. D. (2015). Quantitative analysis for plasma etch modeling using optical emission spectroscopy: Prediction of plasma etch responses. *Industrial Engineering and Management Systems, Korean Institute of Industrial Engineers*, 14(4), 392–400.

36. Li, B., Li, C., Zhao, Y., Han, C., & Zhang, Q. (2020). Deep reactive ion etching of Z-cut alpha quartz for MEMS resonant devices fabrication. *Micromachines (Basel)*, 11(8), 724.

37. Wu, C., Fang, X., Liu, F., Guo, X., Maeda, R., & Jiang, Z. (2020). High speed and low roughness micromachining of silicon carbide by plasma etching aided femtosecond laser processing. *Ceramics International*, 46(13), 17896–17902.

38. Ko, J., Bae, J., Park, M., Jo, Y., Lee, H., Kim, K., Yoo, S., Nam, S. K., Sung, D., & Kim, B. (2023). Computational approach for plasma process optimization combined with deep learning model. *Journal of Physics D: Applied Physics*, 56(34), 344001.

39. Chien, K. C., & Chang, C. H. (2022). Controlling the etch selectivity of silicon using low-RF power HBr reactive ion etching. *Journal of Vacuum Science & Technology B*, 40(6), 062802.

40. Pigatto, L., Baruzzo, M., Bettini, P., Bolzonella, T., Manduchi, G., & Marchiori, G. (2017). Control system optimization in fusion plasmas: The RFX-mod experience. *IEEE Transactions on Nuclear Science*, 64(6), 1420–1425.

41. Cruz, N., Moret, J.-M., Coda, S., Duval, B. P., Le, H. B., & Rodrigues, A. P. (2015). An optimal real-time controller for vertical plasma stabilization. *IEEE Transactions on Nuclear Science*, 62(6), 3126–3133.

42. Bidollahkhani, M., & Kunkel, J. M. (2024). Role of AI in predictive maintenance strategies. *Proceedings of the IARIA, CLOUD COMPUTING 2024: The Fifteenth International Conference on Cloud Computing, GRIDs, and Virtualization*, 49–57

43. Kurokawa, F., Hori, M., Goto, T., Den, S., Katagiri, T., & Kano, H. (2013). Microfabrication of lead-free (K, Na)NbO3 piezoelectric thin films by dry etching. *2013 Transducers & Eurosensors XXVII: The 17th International Conference on Solid-State Sensors, Actuators and Microsystems (TRANSDUCERS & EUROSENSORS XXVII)*, Barcelona, Spain, 1051–1054.

44. Ashraf, M., Sundararajan, S. V., & Grenci, G. (2017). Low-power, low-pressure reactive-ion etching process for silicon etching with vertical and smooth walls for mechano-biology application. *Journal of Micro/Nanolithography, MEMS, and MOEMS*, 16(3), 034501.

45. Dussart, R., Tillocher, T., Lefaucheux, P., & Boufnichel, M. (2014). Plasma cryogenic etching of silicon: From the early days to today's advanced technologies. *Journal of Physics D: Applied Physics*, 47(12), 123001.

46. Choi, G., Efremov, A., Lee, D. K., Cho, C. H., & Kwon, K. H. (2023). On mechanisms to control SiO_2 etching kinetics in low-power reactive-ion etching process using CF_4 + C_4F_8 + Ar + He plasma. *Vacuum*, 216, 112484.
47. Ouhabaz, M., Belharet, D., Micard, Q., Costanza, M., Giuffrida, G., Bartasyte, A., Trigona, C., & Margueron, S. (2024). Microfabrication of piezoelectric MEMS based on thick $LiNbO_3$ single-crystal films. *Nanotechnology*, 35(18), 185504.
48. Ren, Z., Heard, P. J., Marshall, J. M., Thomas, P. A., & Yu, S. (2008). Etching characteristics of $LiNbO_3$ in reactive ion etching and inductively coupled plasma. *Journal of Applied Physics*, 103(3), 034109.
49. Voronin, S., & Vallée, C. (2024). 50 years of reactive ion etching in microelectronics. *IEEE Transactions on Materials for Electron Devices*, 1, 49–63.

8 Industrial Adoption and Future Roadmap for Green MEMS Etching

8.1 INDUSTRY CASE STUDIES IN GREEN MEMS ETCHING

The industrial landscape for microelectromechanical systems (MEMS) fabrication is undergoing a profound shift, driven by the growing imperative for sustainable manufacturing. Among the most environmentally intensive processes in MEMS production is etching, which often involves the use of high-global-warming-potential (high-GWP) gases and hazardous wet chemicals. In response to evolving regulations, supply chain pressures, and environmental, social, and governance (ESG) metrics, several industry leaders have pioneered green etching technologies. This section presents detailed case studies from STMicroelectronics (ST), Bosch, Taiwan Semiconductor Manufacturing Company (TSMC), and emerging green-tech developers, offering a comprehensive look at how these organizations have adopted, scaled, and optimized eco-friendly etching processes in MEMS production.

ST has taken a leadership role in embedding green chemistry into its high-volume MEMS production workflows. According to its 2024 Sustainability Report, the company has committed to fully eliminating SF_6 and CF_4 gases—commonly used for reactive ion etching—by 2027 as part of its roadmap toward carbon neutrality. ST has substituted these gases with argon-hydrogen (Ar/H_2) plasma chemistries, which are inert, non-toxic, and offer significantly lower GWP values. This shift is technically challenging due to the reduced reactivity of Ar/H_2 compared to traditional fluorinated gases. To address this, ST leverages AI-powered plasma control systems and real-time endpoint detection algorithms that dynamically adjust etch conditions to maintain uniformity and critical dimension fidelity. This adaptive control reduces overetching and defect formation while minimizing gas consumption. The integration of these green plasma processes into production has resulted in a reported 42% reduction in GHG emissions per wafer and up to 25% improvement in etch uniformity across sensor platforms such as accelerometers and gyroscopes [1].

A further innovation at ST involves the application of electrochemical etching powered by triboelectric nanogenerators (TENGs). Wang et al. (2024) demonstrated a system in which single electrolyte droplets are electrically activated by mechanical motion, thereby replacing conventional power supplies with energy-harvesting platforms. The system is capable of etching and depositing metal oxides at nanometer resolution, with near-zero chemical waste. TENG-driven electrochemical etching is particularly well suited for localized, application-specific MEMS devices such as biosensors and chemical detectors. ST is exploring this technology in its bioMEMS lines, where it offers advantages in energy savings, process miniaturization, and waste reduction [2].

DOI: 10.1201/9781003674795-8

Bosch, another global MEMS leader, has taken a complementary approach by focusing on green resists and sustainable post-processing. Traditionally, Bosch employed SU-8 photoresists in MEMS lithography, but these resists require toxic solvents and generate large volumes of hazardous waste. In collaboration with academic and industrial partners, Bosch has transitioned to trade mark ADEX® dry film photoresists for its 3D microfabrication processes. ADEX resists are solvent-free, recyclable, and compatible with standard cleanroom lamination tools. Roos et al. (2022) demonstrated that ADEX materials could be used to form robust microchannels and cantilever structures with high yield and resolution, while eliminating VOC emissions and resisting runoff into effluent streams [3].

In Bosch's implementation, ADEX-based fabrication is paired with plasma-based post-etch cleaning using low-energy Ar/O_2 mixtures. This hybrid approach reduces both thermal budget and chemical exposure for device layers. Moreover, Bosch employs AI-enabled digital twins—simulation models trained on historical process data—to optimize the application of ADEX films and plasma parameters in real time. According to internal data, this has improved process consistency and reduced operator intervention by 30%. These green process enhancements have already been applied in Bosch's MEMS platforms for microfluidic diagnostics and indoor air quality sensors [3].

TSMC, while primarily focused on logic and memory fabrication, has also established a sustainable pathway for CMOS-integrated MEMS devices. In a collaborative study led by Yang et al. (2022), a CMOS-MEMS flow sensor was fabricated using the UMC 0.18 μm CMOS process with green post-processing steps. Notably, XeF_2 vapor-phase etching was employed to selectively remove sacrificial silicon layers without damaging underlying metals or oxides. XeF_2 offers a low-temperature, isotropic etch process that does not require plasma or aggressive exhaust handling systems, making it environmentally and operationally favorable [4].

The use of XeF_2 allowed the flow sensor to be post-processed directly on standard CMOS wafers, maintaining compatibility with conventional IC toolsets and reducing the need for additional fabrication infrastructure. The process required minimal chemical waste handling and eliminated the need for aqueous rinsing steps, thus conserving ultrapure water—a major resource concern in semiconductor fabs. TSMC's green etching approach is particularly significant because it demonstrates how large-scale foundries can integrate sustainable post-CMOS processing without disrupting production flows or requiring major capital expenditures [4].

Beyond these industry giants, several emerging technologies are redefining the boundaries of green MEMS fabrication. Wang et al.'s (2024) TENG-based electrochemical platform mentioned earlier represents one such disruptive innovation. In their design, a microneedle-guided droplet containing metal salt electrolyte is actuated across a substrate surface. The voltage generated by the TENG drives localized oxidation or reduction reactions, enabling precise etching or material deposition. The system operates under ambient conditions and requires no toxic solvents, pressurized gas lines, or plasma sources. Its modularity and scalability position it well for lab-on-chip manufacturing, distributed fabrication, and low-resource environments [2].

These advances in sustainable etching are reinforced by broader process lifecycle assessments. Mullen and Morris (2021) emphasized the need to consider not only etching chemistry but also upstream and downstream impacts such as mask material usage, tool energy consumption, and emissions from solvent recovery systems. Their study of green nanofabrication in the semiconductor sector highlighted the benefits of adopting dry film resists, low-energy plasma tools, and closed-loop effluent recycling. When these measures are applied collectively, the environmental footprint of MEMS fabrication can be reduced by more than 50% compared to legacy methods [5].

Artificial intelligence (AI) has emerged as a critical enabler in this transition. According to Podder et al. (2023), AI plays a vital role in MEMS process optimization, particularly in the context of sustainable etching. AI-based models are used to predict etch depth, identify deviations from target profiles, and perform corrective adjustments in plasma density and chamber pressure (Figure 8.1). In Bosch's and ST's green MEMS lines, AI is integrated into both process control loops and

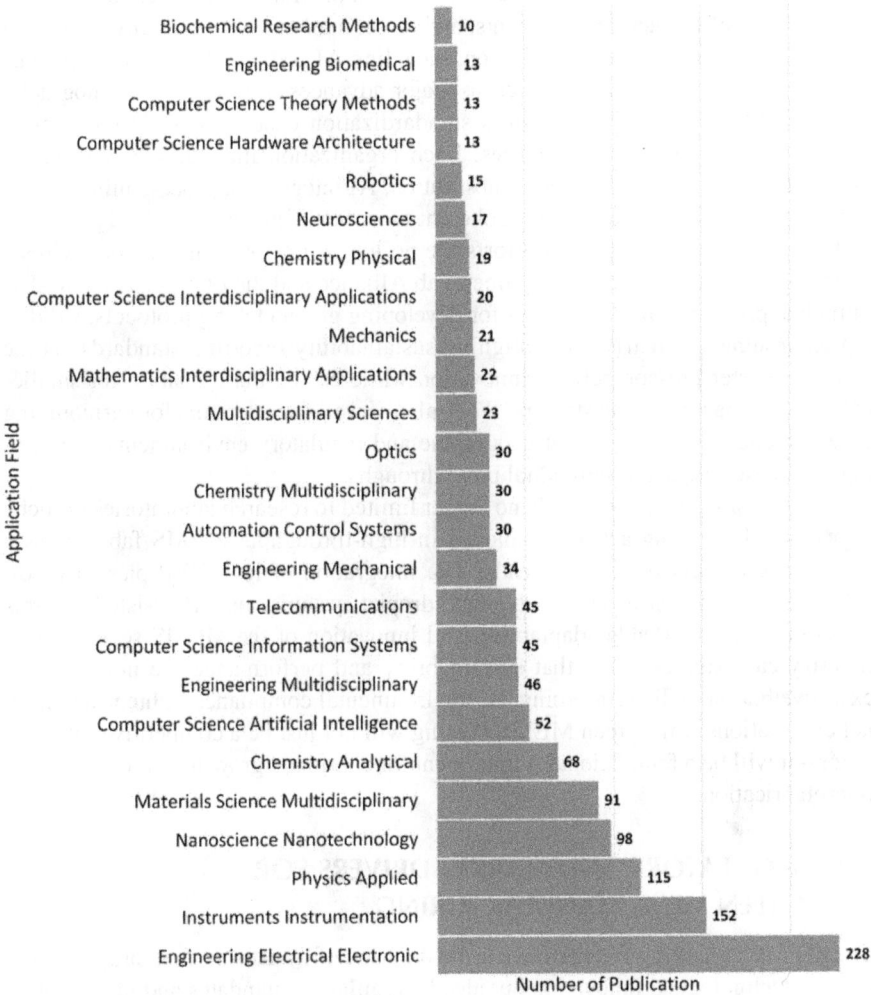

FIGURE 8.1 Combination of AI and MEMS sensor applications in different fields [6].

factory-level resource management systems. This allows predictive maintenance of etching equipment, dynamic gas allocation, and waste stream tracking, all of which contribute to sustainability goals while improving yield and reducing costs [6].

Another notable observation from Podder et al. is that AI and machine learning frameworks are accelerating the adoption of new chemistries that would otherwise require lengthy tuning and validation. In traditional fabs, replacing SF_6 with Ar/H_2 could take months of characterization. With AI-enhanced modeling, this timeline is reduced significantly, allowing faster industrial deployment of environmentally benign gases. The combination of advanced sensors, cloud-based process logging, and edge analytics allows fabs to implement greener workflows without sacrificing precision or repeatability [6].

Taken together, these case studies demonstrate that green MEMS etching is not a one-size-fits-all solution but a diverse ecosystem of strategies tailored to organizational goals, infrastructure constraints, and target markets. ST' vertically integrated model enables deep process innovation, including AI and TENG-powered systems. Bosch's focus on bioMEMS has led to major advances in solvent-free lithography and digital twin simulation. TSMC's standardization enables scalable green post-processing in high-volume foundries. Each organization illustrates how different facets of sustainability—chemical substitution, AI integration, process miniaturization—can be harmonized within a coherent manufacturing vision.

Importantly, these industrial efforts are no longer operating in isolation. Cross-sector collaborations, such as the Green Fab Alliance and the European NANO-EH initiative, provide shared platforms for developing green etching protocols, validating environmental metrics, and aligning sustainability reporting standards. These consortia foster pre-competitive innovation while easing the transition for smaller MEMS manufacturers and startups. They also serve as accelerators for harmonizing green practices across different geographic and regulatory environments, amplifying the overall impact of individual breakthroughs.

In conclusion, green etching is no longer limited to research laboratories or niche pilot lines. It is being actively deployed in high-throughput MEMS fabs across a wide range of devices and materials. The integration of low-GWP plasma gases, solvent-free photoresists, electrochemical droplet systems, and AI-assisted controls illustrates the remarkable adaptability and innovation of the MEMS sector. These industry case studies prove that sustainability and performance are not mutually exclusive but mutually reinforcing. As environmental compliance tightens and market expectations shift, green MEMS etching will not just be a competitive differentiator—it will be a foundational requirement for continued growth and innovation in microfabrication.

8.2 REGULATORY AND MARKET DRIVERS FOR GREEN MEMS MANUFACTURING

The MEMS industry is navigating a critical period of transformation, propelled not only by technological innovation but also by regulatory mandates and market pressures focused on sustainability. Etching processes—among the most material- and energy-intensive steps in MEMS fabrication—are under increasing scrutiny due to their historical reliance on toxic solvents, high-GWP gases, and high water and energy consumption. As environmental compliance becomes a defining component of industrial competitiveness, the transition to green etching is being shaped by a matrix of governmental regulations, sustainability reporting frameworks, and evolving investor expectations. This section explores the regulatory and market landscape influencing the adoption of green MEMS etching technologies and outlines how policy and economic factors are working together to redefine industry standards.

At the core of regulatory attention is the reduction of fluorinated greenhouse gases, particularly sulfur hexafluoride (SF_6), carbon tetrafluoride (CF_4), and nitrogen trifluoride (NF_3), which have global warming potentials thousands of times greater than CO_2. These gases have traditionally been used in deep reactive ion etching (DRIE) and other dry etching processes due to their high selectivity and anisotropy. However, their environmental persistence and climate impact have made them key targets for phase-out initiatives. The European Union's Regulation (EU) No 517/2014, known as the F-Gas Regulation, mandates a staged reduction in the use of these gases across industries, including the semiconductor sector [7]. Similarly, in the United States, the Environmental Protection Agency (EPA), through its Significant New Alternatives Policy (SNAP) program, has classified SF_6 and CF_4 as hazardous substances and encourages their substitution with lower-impact alternatives [3, 8].

These legislative directives have practical consequences for MEMS foundries and integrated device manufacturers (IDMs). Facilities that fail to transition away from high-GWP gases face not only legal risks but also increased operational costs related to gas capture, abatement, and environmental reporting. In anticipation of these regulatory changes, industry leaders are actively substituting conventional etchants with more sustainable options. Blair et al. (2024) demonstrated that indium tin oxide (ITO) meta-surfaces—widely used in MEMS displays and sensors—can be patterned using Ar/H_2 plasma, which produces no greenhouse gases and maintains excellent etch resolution [9]. This process, while requiring precise plasma tuning, provides a feasible pathway for regulatory compliance without compromising on etching performance.

In parallel with regulatory mandates, the MEMS industry is aligning with voluntary but increasingly influential sustainability reporting frameworks. These include the Global Reporting Initiative (GRI), the Sustainability Accounting Standards Board (SASB), and the Task Force on Climate-related Financial Disclosures (TCFD). These standards require companies to quantify and report key environmental metrics, such as Scope 1 and Scope 2 emissions, hazardous waste generation, energy usage, and water intensity. While not legally binding, adherence to these frameworks is rapidly becoming a prerequisite for attracting institutional investment, entering public procurement markets, and securing long-term contracts with ESG-conscious OEMs.

ST provides a leading example of this shift. In its 2024 Sustainability Report, the company disclosed a detailed roadmap to phase out all fluorinated process gases by 2027 and transition to low-GWP etching chemistries across all MEMS fabs [1]. This commitment is supported by real-time gas monitoring systems and AI-enabled process optimization that ensures tighter control over etch profiles and gas flows. In its inertial MEMS product lines, the company reports over 40% reductions in emissions per wafer and improved tool uptime due to reduced plasma-induced contamination. By embedding environmental KPIs into its enterprise-wide performance metrics, ST has aligned its green etching roadmap with corporate governance, investor relations, and marketing strategies.

Bosch, another industrial frontrunner, has integrated sustainability standards into both R&D and operational workflows. In its bioMEMS development programs, Bosch employs ADEX dry film photoresists, which are solvent-free, recyclable, and cleanroom-compatible. These materials reduce chemical waste, improve lithographic consistency, and eliminate hazardous disposal procedures traditionally associated with SU-8 and similar photoresists [3]. Bosch complements this materials shift with digital twins and AI-based process modeling to reduce rework and optimize etch uniformity. These methods, aligned with SASB and GRI standards, have enabled Bosch to not only meet European RoHS and REACH directives but also strengthen its ESG profile for investor transparency.

In Taiwan, TSMC and the broader CMOS-MEMS ecosystem are integrating sustainability into post-CMOS MEMS processing. A 2022 study by Yang et al. illustrated the use of XeF_2 etching in a UMC 0.18 μm CMOS MEMS flow sensor process. XeF_2, a dry vapor-phase etchant, offers high selectivity, room-temperature operation, and a low environmental burden. Its adoption eliminates the need for aggressive fluorinated gases and aqueous rinsing steps, both of which are environmentally costly. The process is fully compatible with existing foundry tools, enabling sustainable MEMS integration within standard CMOS production flows [4].

Beyond compliance and transparency, financial markets and investor sentiment are emerging as powerful accelerators of green MEMS manufacturing. Over the past decade, global sustainable investment has grown to over $35 trillion, and companies with demonstrable ESG commitments now receive preferential access to capital. For MEMS firms, adopting green etching practices improves positioning in ESG ratings and sustainability indices, which increasingly influence procurement decisions and equity valuations. Sustainability-linked loans (SLLs) and green bonds provide further incentives, tying interest rates or capital availability to emissions reductions or circular economy metrics.

Doyle et al. (2023) illustrated the power of such incentives in their work on a low-TRL energy-harvesting MEMS platform. The project, funded in part through the EU Horizon program and NANO-EH initiative, applied the 12 principles of green engineering from concept through fabrication. By eliminating toxic materials, optimizing for disassembly, and using eco-friendly solvents, the team achieved not only sustainability benchmarks but a compelling value proposition for commercialization. The project demonstrates that green MEMS design, including sustainable etching, is not only viable but profitable in the context of modern funding ecosystems [10].

The role of OEMs and end-user markets must also be emphasized. Major electronics firms such as Apple, Samsung, and automotive Tier-1 suppliers now include environmental impact assessments in their MEMS sourcing strategies. This cascading influence pressures MEMS foundries to demonstrate traceability, compliance, and emissions reductions at the process level. In particular, companies supplying components for biomedical, wearable, and automotive applications face customer-mandated reporting on chemical usage, process water management, and embodied carbon. In response, MEMS producers are adopting green etching practices not only for internal cost or compliance reasons but to secure supply chain positioning.

Podder et al. (2023) provide further insight into how AI-driven process control is reinforcing this market adaptation. In their comprehensive review of AI in MEMS manufacturing, the authors highlight how machine learning is used to predict etch rates, monitor endpoint deviations, and manage chamber conditions under fluctuating loads. AI models allow fabs to experiment with green chemistries like Ar/H_2, Cl_2, or XeF_2 while maintaining precise profile control and defect mitigation. In Bosch's and ST's production lines, AI systems are directly linked to sustainability dashboards, enabling real-time tracking of emissions, yield, and tool utilization. These systems allow faster adoption of green processes by minimizing the risks and downtime traditionally associated with chemical transitions [6].

It is also worth noting the influence of national industrial strategies and transnational funding mechanisms. In Europe, the Green Deal Industrial Plan and REPowerEU initiatives offer direct subsidies for fabs adopting low-emission chemistries, upgrading toolsets, or achieving carbon neutrality. In Asia, Japan and South Korea have launched green semiconductor R&D programs, while Taiwan's Ministry of Science and Technology offers grants for sustainable MEMS pilot lines. These programs lower the barrier for smaller MEMS manufacturers to adopt advanced green etching workflows and align with international regulatory harmonization.

In summary, the convergence of regulatory action, sustainability reporting, capital markets, and customer expectations has created a robust framework for the adoption of green etching in MEMS manufacturing. Legislative frameworks like the EU F-Gas Regulation and the EPA SNAP program enforce bottom-line requirements. Voluntary standards like GRI and TCFD provide guidance and legitimacy. Investors and end markets drive strategic alignment. Together, these forces not only incentivize sustainability but make it essential to the operational and financial health of MEMS producers.

The case studies and research insights presented here demonstrate that green MEMS etching is not simply an engineering challenge—it is a multi-dimensional system redesign that touches compliance, economics, and innovation. From ST' AI-controlled plasma chemistries to TSMC's XeF_2-based CMOS integration, green etching is being actively implemented in some of the world's most sophisticated fabs. As regulations tighten and ESG criteria proliferate, these practices will increasingly define the baseline for competitive participation in the global MEMS market.

8.3 AI AND ROBOTICS FOR SUSTAINABLE MEMS MANUFACTURING

The integration of AI and robotics into MEMS fabrication is revolutionizing how manufacturers approach sustainability. Traditional MEMS etching processes have long been plagued by environmental inefficiencies, including the use of high-GWP gases, volatile chemicals, and energy-intensive tools that often suffer from yield loss due to variability in etch performance. As regulatory pressure and ESG investment standards accelerate, AI and robotic automation offer transformative pathways to green MEMS manufacturing. By enabling data-driven optimization, predictive control, and precision robotics, these digital technologies are reducing emissions, waste, and rework in MEMS etching while enhancing performance, reliability, and cost-effectiveness.

AI-driven systems are already being employed across the semiconductor industry to monitor etch processes, predict anomalies, and fine-tune process parameters. In MEMS fabrication, this technology is particularly powerful given the tight geometries, multilayer structures, and high aspect ratios involved. For example, ST has implemented an AI-enhanced plasma etching control system across its MEMS fabs to support the substitution of traditional etching gases like SF_6 and CF_4 with Ar/H_2 plasma—a cleaner but less chemically aggressive alternative. AI models, trained on plasma emissions and endpoint sensor data, dynamically adjust RF power, chamber pressure, and gas flow rates in real time, ensuring uniform etch profiles despite the less reactive nature of Ar/H_2 gas chemistry [1]. This digital control loop compensates for variability in gas dynamics and helps avoid overetching, a common challenge when using low-GWP plasmas.

The importance of AI in scaling sustainable MEMS processing has been thoroughly documented by Podder et al. (2023), who provide a taxonomy of AI applications across MEMS process modules including photolithography, deposition, etching, and packaging. Their findings underscore that etching benefits most from model-predictive control (MPC) systems and supervised learning algorithms, which improve endpoint prediction and pattern fidelity without the need for destructive metrology [6]. In Bosch's MEMS packaging lines, for instance, AI-assisted thermal plasma etching is used in conjunction with ADEX dry film photoresists to minimize chemical exposure and reduce defect rates. The system evaluates dozens of process variables simultaneously—such as chamber wall temperature, mask wear, and back pressure—and then optimizes recipes in real time. According to Bosch's internal data, this approach reduced VOC emissions by over 60% and boosted first-pass yield by 15% [3].

Beyond software, the integration of precision robotics has fundamentally altered how MEMS wafers are handled, aligned, and processed. Robotic arms equipped with vision systems are used to transport wafers through etch chambers, resist stations, and metrology tools with sub-micron alignment precision. This not only improves throughput and repeatability but also significantly reduces cross-contamination and particle generation—two factors that often lead to yield loss and rework. In cleanroom environments, robotic transfer systems also support tool minimization strategies by enabling the consolidation of multiple etch steps into hybrid modules. This reduces total tool count, fab footprint, and energy consumption, all of which are key to achieving green manufacturing metrics [11].

The role of robotics in environmental sustainability is particularly prominent in wet etching and cleaning stages, where human involvement poses contamination and safety risks. In Bosch's hybrid MEMS packaging lines, robots execute complex workflows that integrate solvent-free lamination, dry etching, and green wet cleaning—often using low-toxicity chemicals such as ammonium bifluoride or citric acid. The use of robotics ensures consistent processing conditions and enables the capture and recycling of rinse solutions, leading to more than 40% reduction in water usage and hazardous waste generation [3]. These robotic systems are increasingly being embedded with AI agents that schedule preventive maintenance, optimize tool matching, and perform statistical process control—all without human intervention.

At the edge of innovation, AI and robotics are also being deployed in experimental fabrication platforms such as TENG-powered electrochemical etching. In a study by Wang et al. (2024) [2], a microneedle-controlled robotic arm delivers a microdroplet of electrolyte to a target substrate, powered by mechanical energy harvested from environmental motion. The etching process—governed by a closed-loop AI controller—monitors the local electrochemical conditions and dynamically adjusts needle trajectory, voltage input, and reaction duration to achieve nanometer-scale etching [6]. The entire system operates under ambient conditions, uses no hazardous gases, and requires no cleanroom infrastructure. This self-contained unit exemplifies how AI and robotics can be used not only for automation but for developing entirely new classes of ultra-sustainable MEMS fabrication tools.

In high-volume fabs, the implementation of digital twins has become another powerful method for driving sustainability. A digital twin is a real-time virtual replica of a physical process, used to simulate, monitor, and optimize manufacturing operations. Bosch and ST have developed digital twins of their plasma etching chambers that incorporate real-time sensor inputs, AI modeling, and physics-based simulations. These systems predict etch profile evolution, gas utilization efficiency, and tool degradation. When integrated with fab-wide material tracking systems, digital twins allow early detection of drift or contamination that might otherwise go unnoticed. This helps avoid scrap, improves uptime, and enables predictive scheduling of maintenance and calibration, all of which support more sustainable fab operations [4, 10].

Moreover, AI-assisted design optimization tools are now being used to generate MEMS layouts that are inherently easier to fabricate using green etching processes. For example, topology optimization algorithms can redesign support structures, membranes, and microchannels to reduce the number of etch steps or minimize reliance on high-aspect-ratio features that require aggressive chemistry. This design-for-sustainability approach is being supported by CAD software that incorporates environmental impact assessments into the early design stages. Such methods are helping MEMS designers align with the principles of green engineering, as described by Doyle et al. (2023), where performance is balanced with process sustainability and end-of-life considerations [10].

The broader implication of these technologies is that sustainability in MEMS etching is not limited to material substitution or energy reduction alone. It is being driven by a systemic transition toward cognitive manufacturing—a paradigm where data, machines, and algorithms cooperate to achieve optimal outcomes with minimal waste. AI and robotics are not only replacing manual labor and hard-coded recipes; they are redefining how fabs think, learn, and improve over time. In this paradigm, green etching becomes not an isolated goal but a core metric of intelligent manufacturing.

It is also important to consider how AI and robotics enable regulatory compliance and reporting. With increasing requirements from frameworks like GRI, SASB, and TCFD, fabs must now provide detailed and auditable data on emissions, energy usage, chemical inventories, and process waste. AI systems can automate the extraction and formatting of this data from tool controllers, sensors, and manufacturing execution systems (MES), drastically reducing the reporting burden while improving accuracy. These digital records support not only compliance but also continuous improvement, as they highlight areas where waste can be reduced or efficiency improved [12].

Financial incentives are further amplifying the adoption of AI and robotics in sustainable MEMS fabrication. Investors and public procurement agencies increasingly assess ESG performance as a factor in decision-making. Companies that demonstrate leadership in smart, green manufacturing are rewarded with better financing terms, higher stock ratings, and access to SLLs. ST and Bosch have already leveraged their AI-enabled green etching platforms as part of investor communications and public sustainability reports, thereby reinforcing their market leadership and securing funding for further innovation [1, 3].

In conclusion, AI and robotics are not auxiliary tools in the quest for sustainable MEMS manufacturing—they are fundamental enablers. Through process modeling, predictive control, design automation, and robotic execution, these technologies allow MEMS manufacturers to transition away from high-emission etching processes while improving yield, reliability, and compliance. From cleanroom fabs to decentralized electrochemical etching stations, intelligent systems are reshaping the MEMS production landscape for the better. As sustainability expectations continue to rise, the industry's ability to integrate AI and robotics into its green etching workflows will determine not only its regulatory compliance but also its global competitiveness and long-term viability.

8.4 STRATEGIC PARTNERSHIPS AND ROADMAP ALIGNMENT

The transition toward green MEMS etching requires more than technological substitution—it necessitates a collaborative realignment of supply chains, knowledge networks, and regulatory frameworks. As environmental standards tighten globally and MEMS applications grow increasingly complex, strategic partnerships have emerged as the cornerstone of sustainable manufacturing transformation. Industry leaders are no longer operating in silos; instead, they are forming multi-stakeholder alliances involving academia, foundries, equipment suppliers, and governmental bodies to develop and deploy green etching innovations. These partnerships are shaping the roadmap for sustainable MEMS at both the micro (tool-level) and macro (policy and infrastructure) scales.

One of the clearest indicators of this shift is the emergence of collaborative pilot lines that focus explicitly on eco-friendly fabrication methods. For example, the NANO-EH initiative—a European Union-supported platform—unites academic researchers, small-to-medium enterprises (SMEs), and major MEMS companies in a consortium that applies the 12 Principles of Green Engineering to the full MEMS development pipeline. Doyle et al. (2023), one of the key participants, describe how the initiative enabled the development of energy-harvesting MEMS devices using solvent-free lithography, recyclable substrates, and disassembly-friendly architecture. Importantly, these practices were not developed in isolation but were validated in shared facilities with joint oversight from process engineers and environmental regulators [10]. This underscores the value of shared technical infrastructure in advancing and standardizing green etching protocols.

Collaborations are also vital for the translation of academic research into production-ready processes. Many emerging green etching techniques—such as TENG-driven electrochemical etching or laser-assisted dry etching—remain at low technology readiness levels (TRLs). Without industrial partnerships, these innovations often remain trapped in lab-scale demonstration. For example, Wang et al. (2024) developed a TENG-powered single-droplet etching system capable of nanometer-resolution patterning using ambient electrochemistry. This method, powered by mechanically generated voltage, eliminates the need for external power supplies and cleanroom tools, representing a leap toward ultra-sustainable MEMS processing [2]. However, industrial deployment requires joint development agreements (JDAs) with firms that can adapt these concepts to wafer-scale platforms. Public-private partnerships (PPPs), particularly in regions with strong R&D tax incentives, offer fertile ground for such scale-up initiatives.

Foundry ecosystems provide another crucial mechanism for roadmap alignment. The case of TSMC and UMC in Taiwan illustrates how green MEMS etching can be embedded into the foundry model. In their 2022 paper, Yang et al. detail how a CMOS-MEMS flow sensor was fabricated using standard CMOS processes and a post-CMOS green XeF_2 etch step that minimized chemical waste and plasma-induced damage. This integration was only possible because the foundry offered a modular process library with validated green post-processing modules [4]. By providing clients with environmentally optimized process options, the foundry acts as a conduit for technology diffusion. This model demonstrates how roadmap alignment can be achieved through platform modularity, reducing the need for each designer to reinvent environmental solutions from scratch.

Roadmap alignment is further supported by industry-wide standardization efforts. Bodies such as SEMI, IEEE, and the International Roadmap for Devices and Systems (IRDS) are beginning to incorporate sustainability criteria into their process standards. The IRDS 2024 roadmap explicitly highlights green etching as a priority area, citing the need for low-GWP plasma gases, solvent-free resists, and minimal water footprints. This marks a departure from traditional roadmapping, which focused primarily on performance and cost metrics. With sustainability now a first-order concern, these standards are being reshaped by interdisciplinary working groups composed of engineers, environmental scientists, and policy advisors.

Governmental and intergovernmental organizations are playing a complementary role by funding roadmap-aligned pilot projects. In Europe, the Green Deal Industrial Plan and Horizon Europe programs offer substantial grants and tax benefits to fabs that demonstrate carbon reduction via process innovation. Similarly, Taiwan's Ministry of Economic Affairs has launched green semiconductor initiatives that incentivize fabs to reduce chemical use, adopt closed-loop water systems, and report Scope 1 and 2 emissions data. These policy levers do not operate in isolation—they are often tied to membership in international collaborations such as the Global Semiconductor Alliance (GSA) and the Asia-Pacific Green Fab Consortium, which facilitate benchmarking, knowledge sharing, and collective negotiation with tool vendors.

Tool vendors and material suppliers are essential partners in this ecosystem. Their role in developing green etching hardware—such as low-GWP gas delivery modules, dry film laminators, and real-time emissions monitoring systems—is foundational to industrial adoption. For instance, suppliers of ADEX dry film photoresists have worked directly with MEMS fabs to tailor resist thickness, adhesion promoters, and etch resistance to meet the stringent demands of microfluidic and cantilever applications [3]. These collaborations go beyond supplier-customer relationships, evolving into co-development models with shared intellectual property, pilot-line access, and rapid feedback loops. Such models align with the "triple helix" framework of innovation, wherein academia, industry, and government co-create and validate sustainable technologies.

AI and digital infrastructure also serve as alignment mechanisms between partners. As described by Podder et al. (2023), AI-driven data analytics platforms are being shared across fab ecosystems to monitor tool health, gas usage, emissions, and process drift. These systems allow different fabs—even across regions and regulatory regimes—to compare sustainability metrics in real time and align their internal practices with external benchmarks [6]. This creates a level of operational transparency that is essential for cross-border roadmap coordination. For instance, fabs in Europe and East Asia can agree on shared definitions of "green etch readiness," facilitating supplier certification and process migration.

Robotic integration contributes as well to roadmap synchronization. At Bosch and ST, robotic wafer handling systems are deployed not just for productivity but to enforce process uniformity across parallel etch lines. These systems reduce operator variability and enable "copy-exact" etch modules that can be scaled to new sites or adapted to new materials. In TSMC's context, robotic integration also supports the development of modular fab blueprints that include space, utility, and automation specifications for green etching tools. These blueprints can be licensed to partners, accelerating time-to-market while embedding sustainability by design.

Strategic partnerships also mitigate risk and accelerate time-to-yield. For example, the EU-funded REMAPS initiative brought together MEMS designers, etch tool vendors, and regulatory experts to develop a shared risk model for adopting new etch chemistries. The initiative produced open-source guidelines on chemical substitution, emissions quantification, and sustainability-linked performance metrics. These tools help de-risk investment in green technologies by providing evidence-based projections of cost, yield, and regulatory trajectory. Such frameworks are especially valuable for SMEs, who often lack the resources to independently validate green etching methods or absorb the risks of first-time implementation.

As these examples demonstrate, roadmap alignment and strategic partnerships are no longer ancillary strategies—they are prerequisites for sustainable MEMS innovation. The complexity of modern green etching technologies, from hybrid wet-dry schemes to AI-tuned plasma chemistries, demands cross-disciplinary collaboration and institutional coordination. Without shared tools, common standards, and cooperative funding mechanisms, the adoption curve would be prohibitively slow and fragmented.

In conclusion, the future of green MEMS manufacturing will be shaped not only by the capabilities of individual fabs but by the strength of their partnerships and the clarity of their shared roadmaps. Foundries, OEMs, toolmakers, universities, and policymakers must work in concert to create an enabling ecosystem. Strategic collaborations—whether through JDAs, government-funded pilot lines, or international consortia—are critical to scaling green etching methods from concept to industrial routine. The industry must now move beyond proof-of-concept and toward proof-of-alignment, ensuring that sustainability is embedded not just in tools and materials but in the very structure of MEMS innovation.

REFERENCES

1. STMicroelectronics. (2024). *ST sustainability report 2024*. Retrieved from https://www.st.com/company-reports
2. Wang, R. C., Zhou, Y. H., Lee, Y. H., Chen, H. C., & Chen, C. E. (2024). TENG-driven single-droplet green electrochemical etching and deposition for chemical sensing applications. *Applied Surface Science Advances*, 23, 100634.
3. Roos, M. M., Winkler, A., Nilsen, M., Menzel, S. B., & Strehle, S. (2022). Towards green 3D-microfabrication of bio-MEMS devices using ADEX dry film photoresists. *International Journal of Precision Engineering and Manufacturing-Green Technology*, 9(1), 43–57.
4. Yang, L. J., Waikhom, R., Shih, H. Y., & Lee, Y. K. (2022). Foundry service of CMOS MEMS processes and the case study of the flow sensor. *Processes*, 10(7), 1280.
5. Mullen, E., & Morris, M. A. (2021). Green nanofabrication opportunities in the semiconductor industry: A life cycle perspective. *Nanomaterials*, 11(5), 1085.
6. Podder, I., Fischl, T., & Bub, U. (2023). Artificial intelligence applications for MEMS-based sensors and manufacturing process optimization. *Telecom*, 4(1), 165–197.
7. European Commission. (2014). *Regulation (EU) No 517/2014 on fluorinated greenhouse gases*. Retrieved from https://eur-lex.europa.eu/eli/reg/2014/517/oj/eng
8. United States Environmental Protection Agency (EPA). (2021). *Significant new alternatives policy (SNAP) program*. Retrieved from https://www.epa.gov/snap
9. Blair, S. F. J., et al. (2024). Green etching of indium tin oxide metasurfaces. *Optical Materials Express*, 14(7), 1924–1936.
10. Doyle, L., Cavero, G., & Modreanu, M. (2023). Applying the 12 principles of green engineering in low TRL electronics. *Sustainability*, 15(14), 11227.
11. Choi, K., et al. (2024). Eco-friendly glass wet etching for MEMS application: A review. *Journal of the American Ceramic Society*, 107(4), 6497–6515.
12. European Commission. (2021). *Sustainable finance disclosure regulation (SFDR)*. Retrieved from https://finance.ec.europa.eu

Index

A

Abatement, 1–4, 21, 111
Acid-based wet etching, 63, 64, 65
Adsorption (Langmuir model), 14, 15, 17, 18, 30
AI-driven, 11, 12, 46, 54, 56, 58, 60, 95, 99, 113, 114, 118
Anisotropic etching, 15, 23, 29, 35, 57, 71, 79, 82, 90
Anisotropy, 1, 8, 15, 16, 19, 20, 25, 28, 30, 34, 35, 37, 52, 54, 55, 57, 79, 86, 91, 95–97, 101, 111
Atomic Layer Etching (ALE), 11, 45, 41–43, 59, 62, 102, 103
Aspect Ratio Dependent Etching (ARDE), 28, 94, 96, 97

B

BioMEMS, 14, 19, 39, 43, 51, 64, 66, 71, 73, 107, 110, 112
Bosch process, 8, 20, 28, 29
Bromine-based etching (Br_2, HBr), 10, 36–38
Bubble formation, 22, 23, 29
By-products, 2, 8–9, 20, 34–36, 38–40, 43, 54–55

C

Chlorine-based etching, 34
Citric acid, 63, 66, 73–74, 114
Convolutional Neural Networks (CNN), 59
Corrosion, 11, 34, 36–38, 72, 83, 88
Cryogenic, 12, 28, 30, 32, 92, 95, 101, 103, 105

D

Deep Reactive Ion Etching (DRIE), 1, 20, 30, 32, 33, 52, 59, 62, 79, 92, 105, 111
Directionality, 14–16, 20–21, 28, 41, 48, 58, 96
Dry etching, 1, 3, 8, 11, 13, 15–16, 19, 20, 22, 23, 28, 29, 30, 34–39, 41, 43–47, 49, 51, 53–62, 68–69, 74, 77, 79, 86, 87, 90, 102, 104–105, 111, 114, 117

E

Edge damage, 48
Electrochemical Etching (ECE), 22–23, 31, 69–73, 75–77, 79, 81–83, 85, 87–89, 107, 115–117, 119
Endpoint detection, 10–11, 13, 56–57, 59, 62, 98, 107

E

Energy consumption, 1, 3–6, 11–12, 48, 53, 58, 76, 78, 99–100, 109, 111, 114
Environmental impact, 1, 7, 11, 13, 19, 63, 65, 73, 80, 81, 83, 85–87, 112, 115
Etch rate, 8, 10, 11, 14,–15, 17–22, 25, 28–30, 32–35, 39–42, 48, 54–55, 57, 59, 63–66, 69–70, 74, 76, 79, 85–87, 91, 93–94, 97, 99–100, 102, 104, 113
Etch selectivity, 9, 18, 28, 29, 52, 63, 66, 96–97, 100, 101, 105

F

Fluorinated gases, 1–3, 7, 10, 21, 107, 112
Fluorine-free, 7, 10, 11, 13, 19, 21–22, 34–35, 37, 39, 41, 43, 45–47
Femtosecond laser, 32, 49–54, 61, 87, 89, 98, 105

G

Global Warming Potential, 1, 12, 21, 107, 111
Green manufacturing, 12, 69, 81, 114, 116
Greenhouse gas emissions, 5, 8, 86

H

Hazardous, 1–3, 5–7 11, 63, 65, 68–71, 73, 76, 81, 84–85, 107–108, 111–112, 115
Hybrid, 8–12, 16, 18, 22, 32, 36, 46, 50–54, 56, 64–66, 71, 73–74, 82, 85–88, 94, 98–100, 103, 108, 114, 118
Hydrofluoric Acid, 2, 17, 63, 74, 76

I

Ion Beam Etching (IBE), 23, 25, 26 31–32, 46–48, 54, 58, 61
Ionic Liquid, 9–10, 69–70, 75, 81, 82, 84–85, 89
Isotropic etching, 15, 42, 64, 66, 80, 82, 85, 90

K

KOH (Potassium Hydroxide), 2, 23, 29, 30, 76, 77, 79

L

Lactic acid, 63, 64, 66
Langmuir adsorption model, 18
Laser-assisted, 26–27, 32, 52, 65, 117
Life-cycle assessment (LCA), 2, 4, 6–7, 12, 14, 16
Long short-term memory (LSTM), 11, 13, 56, 62

For Product Safety Concerns and Information please contact our EU
representative GPSR@taylorandfrancis.com
Taylor & Francis Verlag GmbH, Kaufingerstraße 24, 80331 München, Germany

www.ingramcontent.com/pod-product-compliance
Lightning Source LLC
Chambersburg PA
CBHW032003180326
41458CB00006B/1714

* 9 7 8 1 0 4 1 1 4 4 9 6 0 *